Radiation Exposures

The hidden story of the health hazards behind official 'safety' standards

LES DALTON

Scribe Publications

First published in Australia in 1991
Reprinted July 1991
Reprinted September 1991

Scribe Publications Pty Ltd
RMB 3120, Lancefield Road
Newham Vic 3442

Published with the co-operation of the
Community Education Publication Association
Inc. Victoria

Output in 11 on 13pt Century Oldstyle
by Town and Country Typesetters Pty Ltd

Printed in Australia by Australian Print Group

National Library of Australia
Cataloguing-in-Publication data:

Dalton, Les, K
Radiation exposures.

Bibliography.
Includes Index.
ISBN 0 908011 19 9.

1. Public health. 2. Radiation — Toxicology.
1. Title.

362.1969897

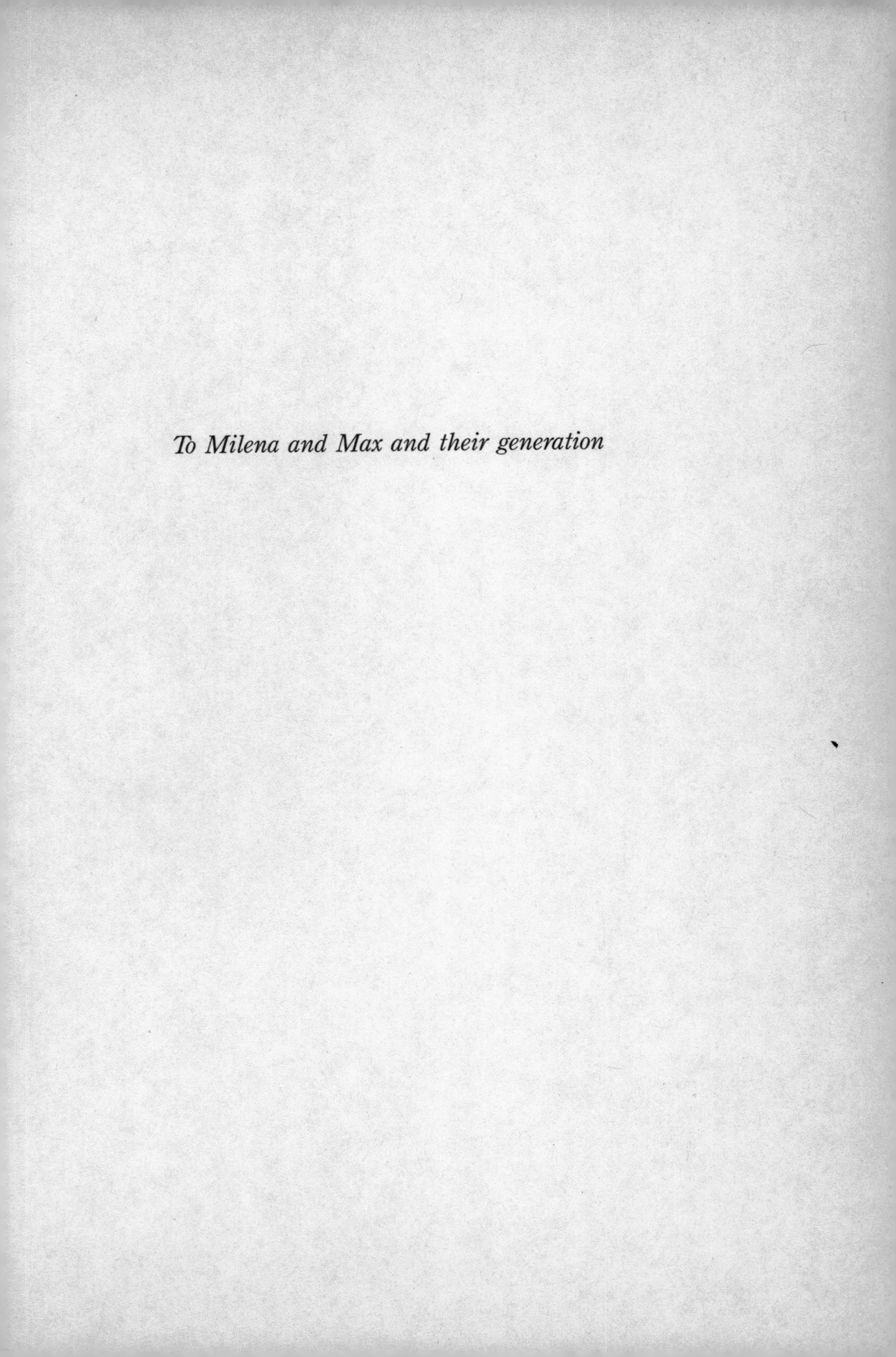

To Milena and Max and their generation

Contents

Acknowledgements

THANKS are due to a number of people who, in one way or another, have helped me in the writing of this book. Many people read drafts, asked questions and suggested changes. I am grateful to Rosaleen Love, who received an early draft from out of the blue, for her thorough reading and suggested changes; the outcome, I am sure, is a more pertinent and understandable text. Other people I would like to thank for this help are: Denise Chevalier, Dorothy Dalton, John Ellis, Pat Jessen, Bert King, Eric Macarow and Gavan Thomson. When the manuscript was finally submitted for publication, Frans Timmermann edited it and further improved its clarity. Thanks also to John Ellis, Peter Green, Harold Haig and Salvatore Rotin for preparation of the figures.

People who have campaigned to limit radiation exposure of their neighbourhoods have also contributed to this book. With some I have been able to discuss the issues personally. John Moore was involved with other residents in stopping the Merri Creek power line, and Leah Healy tried unsuccessfully, along with other South Melbourne residents, to stop the installation of an earth satellite station. Heather Rice has campaigned tirelessly to have the nuclear reactors at Lucas Heights closed down. I have been helped by John Evans, among others, to obtain accounts of the harm done to nuclear veterans and Aborigines, and of their battles for recognition of their injuries. Tape-recordings by Eric Miller and Ila Marks of people's experiences with radiation sources have provided helpful background.

Sandy Doull helped me with his insights into the problems of radiation-related health problems in industry, and the obstacles to improving radiation safety standards encountered by health and safety officers in trade unions. He provided a wealth of information, some of which formed the basis of chapter 5. Alan Roberts gave me the benefit of his knowledge of radiation physics, correcting and refining my elementary presentation of the subject; and Gerry Harrant helped with his admirable familiarity with electronic equipment. Gareth Clayton refined my simplifications of how radiation risk is dealt with statistically.

Listening to Jean McSorley talk about the radiation environment created by the Sellafield reprocessing plant, when she visited Australia in 1988, made me feel more positively that a book like this was needed in order that Australians should hear her warning against an 'Australian Sellafield'.

The scientific information in this book has come from many sources, also with the aid of many people. Knowing a librarian such as Jill Warneke, so willing to search out references, was a great help. So too were the frequent discussions I had with Ian McMillan, and the literature surveys he made while working on the issue of the health effects of power-line fields at the Collingwood Community Health Centre. Though I have criticised the usual responses of our radiation protection bodies to community concerns, I value the scientific quality of their many reports on radiation problems in Australia. Hugh Hamersley, librarian at the Australian Radiation Laboratories, was especially helpful in providing information.

Many accounts of the performance of the nuclear industry and events associated with the industry have come from the national and international information networks that have grown up among social and environmental community groups in Australia and around the world. These alternative networks are loosely structured and rely largely on voluntary effort, but I have found their scientific abstracts and news reports on radiation issues to be consistently reliable. In this regard, I would like, especially, to thank those in Amsterdam who prepare the *News Communiqués* of the World Information Service on Energy.

Finally, I would like to thank Pat Jessen for the way she encouraged me in the belief that, amongst today's rising flood of information, there was a place for a community-oriented book on the social and environmental impact of technological developments.

— *Les Dalton, Melbourne, March 1991*

Abbreviations

AAEC Australian Atomic Energy Commission (now the Australian Nuclear Science and Technology Organisation and the Nuclear Safety Bureau)

ANSTO Australian Nuclear Science and Technology Organisation

ARL Australian Radiation Laboratory

BEIR Committee on the Biological Effects of Ionising Radiation (US)

CSIRO Commonwealth Scientific and Industrial Research Organisation

ICRP International Commission on Radiological Protection

IRPA International Radiation Protection Association

NHMRC National Health & Medical Research Council (Australia)

NRPB National Radiological Protection Board (UK)

NSB Nuclear Safety Bureau

SAA Standards Association of Australia

SECV State Electricity Commission of Victoria

UNSCEAR United Nations Scientific Committee on Effects of Atomic Radiation

USAEC United States Atomic Energy Commission (whose functions are now carried out by the US Department of Energy and the Nuclear Regulatory Commission)

Boxes

Figures

Illustrations

Tables

Preface

WHEN communities first become aware of an external hazard they suspect may be affecting their health, they look to public health organisations for help. Whether the problem is environmental pollution or the threat of infectious micro-organisms, the community expects and usually gets prompt and efficient protective action. However, if the pollutant is not tangible — for example, if it is not smelly or sooty — the official response may be bland; or in the case of unsensed radiation, it may be blandly reassuring.

On such occasions, well-meaning health officials may review the conflicting findings in the relevant research fields, and may decide that there is no scientific basis for suspecting a health hazard. In the case of radiation, a simple statement may be made that the levels of exposure in the community are 'within the acceptable limits set by international safety standards'. But permissible radiation exposure limits have been and still are being revised drastically downwards. Can we trust standards that are constantly being revised?

Obviously, more than science is involved. The politics of environmental pollution are not those of infectious disease. For a start, different economic forces are at work. Governments are reluctant to hamper projects sustaining economic growth or contributing to national security. With a few brave exceptions, health officials faced with evidence of pollution have been inclined to smooth over rather than address the concerns of exposed communities.

At present, communities which perceive a radiation hazard fail not only to get the assistance of bodies that are supposedly responsible for protecting their health, but can find themselves in conflict with them. This is, to say the least, an unfortunate situation.

For example, when the Victorian State Electricity Commission proposed building a high-voltage power line in the 1980s, the project met stiff opposition from local communities. Many objections were raised, and not only because of the obvious unsightliness of power lines. People had learned of likely adverse health effects resulting from the invisible electric and magnetic fields radiating from the lines.

In the face of determined communal opposition, the government appointed an American expert to examine 'the scientific basis of community concerns'. However, many health-related questions about power-line radiation remained unasked and unanswered; some crucial questions are still not being addressed in current research efforts.

Epidemiological studies have shown that children living in power-line electromagnetic fields are more likely than other children to suffer leukaemia. These findings are hotly debated inside the scientific community. The appointment of the expert only further polarised the public debate over the power line's safety.

Community activists had done their homework on the scientific issues and the personalities involved. The expert's findings, when published, were a predictable mixture of fact and personal judgement unacceptable to the community; they did nothing to resolve the political impasse. Eventually, the government was obliged to change its plans to accommodate the community's concerns.

In recent years, battles over power lines have occurred many times around the world. Other battles are being fought over other radiation sources, in situations where people feel threatened by exposure to them.

RADIATIONS come from many different sources and are of several forms. Some are life-giving. Solar radiation, for example, nurtures all life on earth. Others can be life-threatening, according to how they are used or misused. Radiations are broadly classed as either *ionising* or *non-ionising* according to how they behave in matter, including living tissue. Ionising radiations include X-rays and other rays emitted by radioactive material. Non-ionising radiations are emitted by electronic equipment such as that used in telecommunications and power lines, and are usually referred to as *electromagnetic radiation* or simply EMR.

Another way to regard the kinds of impact radiations have on the environment and on human health is to see them as stemming from nuclear and non-nuclear radiation sources. Non-nuclear sources — such as X-ray machines, microwave heating devices, or radar dishes — cease to emit radiation once they are switched off. But once a radioactive source is created, its radiation emissions persist according to physical laws beyond human control. Plutonium, for instance, will emit its highly toxic radiation inexorably for a quarter of a million years. Other radioactive products of the nuclear industry will emit radiation for shorter or even longer times.

Ionising radiations are powerful *mutagens* — that is, they can bring about changes in the way cells reproduce. There is no such thing as a safe dose of ionising radiation. Even extremely low doses can subtly change living cells. Yet health authorities often talk about 'permissible' exposures to radioactive sources that are 'too low' to cause health damage.

It is difficult for people to come to terms with the health hazards of unsensed radiation. During the nuclear weapons testing of the 1950s there was great public concern about radioactive fallout; later, the concern died away, and it took the dramatic events of the Chernobyl nuclear reactor disaster in the Soviet Union to reawaken people to the dangers of radioactivity in food and drinking water.

For communities living on the doorstep of nuclear plants, such as the one treating radioactive waste at Sellafield in Britain, it has been hard to forget. 'You don't lie on the beach anymore or eat the fish', a nearby villager has said. If people try to forget their nuclear neighbour there is always the poignant reminder of a high incidence of leukaemia among the village children.

Evidence of damaged health from non-ionising EMR, such as from power lines, has appeared mostly in the past two decades. The commonplace nature and everyday familiarity of these sources has no doubt made it difficult for ordinary suburban residents to accept the possibility of potential health hazards. However, studies already suggest an excess of cancer among workers in electrical and electronic workplaces.

Heightened concern about health effects from changing environmental factors, such as radiation, is no doubt being influenced by changing sensitivity — some scientists would say excessive sensitivity — about the quality of the environment. We are becoming more aware of ecology and are starting to think more holistically about our health. The charter of the World Health Organisation describes health as: 'A state of complete physical, mental and social well-being'. Health is much more than an absence of disease. Yet the setters of health and safety standards persist in basing their standards on the statistics of mortality. The loss of well-being is not so amenable to statistical treatment as death, and so does not play a part in their thinking.

Of course we have the scope, at least in the affluent world, to adjust our lifestyle voluntarily in order to avoid harmful consumer products and dietary habits; we can perhaps even choose a pollution-free habitat. We can limit our exposure to medical radiations to what we believe is essential to benefit our health. We can avoid excessive exposure to the sun's ultraviolet rays. We might even be choosy about slightly radioactive mineral waters; be careful to vent the radioactive gas, radon from our brick house; and buy a computer monitor that shields us from its EMR emissions. Even so, our choices must rely on safety standards.

But try as we might to adjust our lifestyle there is a limit to what we can achieve personally. Once inside a medical institution, for instance, it is all too easy to be overawed by the professionals and to meekly accept unproductive X-rays. As well, many of our exposures are involuntary. We can do nothing about escaping radioactive particles once they are blown by winds from nuclear explosions or nuclear reactor accidents, or carried on the ocean currents from waste dumps. Europeans cannot escape the radioactive pollutants from Chernobyl now impregnating their soils, nor can the Polynesians avoid the radioactive contamination of their fish from French nuclear explosions. If we live near a power line we cannot, by our own devices, protect our dwelling against pervasive EMR.

THE record of the national and international expert committees responsible for advising on radiation-related health problems gives ground for public disquiet. These bodies did not raise their voice during the nuclear weapons testing of the 1950s and 1960s, which by the end of this century could cause over half a million people to die prematurely. Instead, the experts were busily advising the bomb-testers.

When civilian victims of the fallout in the United States sought compensation they were rebuffed by national health and radiation protection bodies. The courts found that communities had been harmed by the fallout but that, because the tests were for national security, they were not entitled to compensation.

Veterans of nuclear testing at Maralinga have had to battle with the Australian Ionising Radiation Advisory Council over related health injuries. Their radiation experiences were the subject of an expert review that excluded all evidence of lay eyewitness accounts. The benefit of the doubt raised by conflicting scientific evidence on radiation-related health injury has gone mostly to the military.

In the aftermath of Chernobyl, European officials confused people about fallout hazards in order to shield their own national nuclear industries. Now these same authorities are recommending standards with *increased* limits for radioactivity in food, in readiness for the 'next Chernobyl'.

The International Commission on Radiological Protection and other international and national ionising radiation advisory bodies are all self-elective. Their memberships are drawn from among scientists with career interests in the regulation of radiation exposures compatible with the economic goals of nuclear, radiological, and associated activities. The International Radiation Protection Association, which recommends safety standards for non-ionising EMR, is similarly career-orientated.

Insofar as the issues of health in the changing radiation environment are scientific and medical, the public is entitled to be provided with all available information. However, industries and governments in a hurry to see progress made with new technologies or simply anxious to ensure economic growth, fear that, if given the choice, ordinary people will lean towards accepting any positive findings of health hazards as a matter of prudence. As it happens, this instinctive response to potential health hazards is sound, and should be the basis of public health policy.

THIS book attempts to insert clear and relevant information into the radiation debate. Parts I and II give accounts of some social and environmental impacts of the developing radiation industries, including the experiences of affected communities and individuals.

The book's structure is based on a division which has been made between nuclear and non-nuclear radiation sources, because they create distinctly different problems for environmental protection and so for public health policy. The emissions from electronic and electrical installations — the non-nuclear radiations — are dealt with in Part I. Emissions from radioactive substances — the nuclear radiations — are dealt with in Part II.

Part III is for readers who want more detailed information about the scientific basis of radiation-related biological changes and their associated health effects.

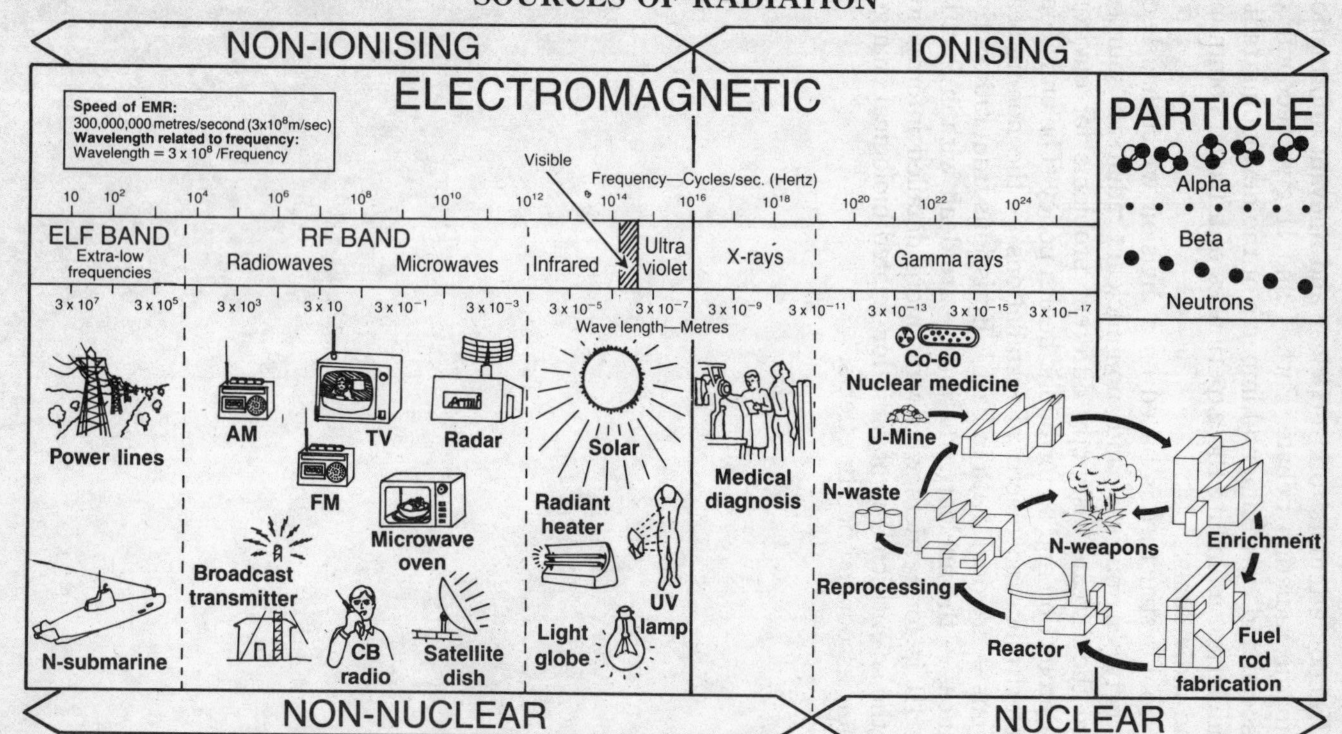

A chart showing how electromagnetic and particle radiations are associated with various medical, telecommunications, nuclear and other activities.

PART I

Non-Nuclear Sources of Radiation

Artificial radiations are emitted as a result of the movement of electrons in electronic and electrical equipment. They are electromagnetic radiations (EMR), such as X-rays, microwaves, and radio waves, as well as the extra-low-frequency electric and magnetic fields from power lines. These radiations from *non-nuclear* sources create different environmental problems from those emanating from *radioactive* materials. Except for X-rays, they also behave differently in living cells.

The harmful effects of X-rays, which are *ionising*, have been known for a long time. There is now evidence that chronic exposure to other artificial EMR associated with telecommunications and power-line emissions can harm human health. These are the *non-ionising* EMR.

However, while much has been learned in recent years about biological changes caused by non-ionising EMR, less is known about the health effects. In the face of such uncertainty about possible health effects, communities are asking that their exposure to non-ionising EMR should be minimised. Much can be achieved to minimise exposure in the workplace and the wider environment, by carefully locating and shielding the EMR sources.

PART I

Non-Nuclear Sources of Radiation

Artificial radiations are emitted as a result of the movement of electrons in electronic and electrical components. They include electromagnetic radiations (EMR) such as [illegible] microwaves and radio waves, as well as the extra low frequency electric and magnetic fields from power lines. These radiations from non-nuclear sources create different environmental problems from those emanating from radioactive materials, because [illegible] they behave differently in living cells.

The harmful effects of X-rays, which are ionizing, have been known for a long time. There is now evidence that chronic exposure to other artificial EMR is associated with [illegible] and [illegible] by non-ionizing EMR.

However, while much has been learned in recent years about the biological changes caused by non-ionizing EMR, less is known about the health effects. In the face of such uncertainty [illegible] exposure to non-ionizing EMR should be minimised [illegible] by [illegible] and [illegible] sources.

Chapter 1

X-rays — use and abuse

> What bids fair to be a most useful addition to our methods in diagnosis has lately been discovered by Professor Roentgen of the Wurtzburg University, Bavaria.
>
> — *F. W. Elsner,* Australasian Medical Gazette, *p. 101, 20 March 1896.*

> Roentgen rays are arguably the major diagnostic tool of the medical profession, second only to the diagnostic acumen of a skilled and experienced physician . . . [They] are also a pathogen, causing cancer in this generation and defects in future ones.'
>
> — *James Wright, former director of pathology at Sydney Hospital,* New Doctor, *p. 18, November 1988.*

ON the winter morning of Sunday, 5 January 1896, people in Vienna opened their copies of the *Presse* to read about the sensational discovery by a German physicist, Wilhelm Roentgen, of powerfully penetrating, yet invisible rays. The rays could be beamed through the human body to take shadow pictures of the bones.

Although Roentgen knew his discovery to be an important breakthrough, he had not sought publicity for it. Indeed, he was drawn reluctantly into the limelight. He had done no more than was his custom. He had submitted a paper entitled 'On a New Type of Rays' to the local Physical Medical Society.[1] Because he valued the esteem of his peers and wanted to contribute to his nation's prestige in the rapidly growing and competitive international scientific fraternity, he had also sent off his pictures to scientific colleagues around Europe. One of his Austrian colleagues leaked the discovery to the *Presse* editor, who immediately realised he had a sensational story.

Within days, stories about the new rays were printed in many countries. Australians received the news weeks later, and then only in a report on how the rays had been used in Britain. Such were the colonial ties.

The name given by Roentgen — X-rays — added a touch of mystery. He rejected financial inducements to develop his X-ray machine for commercial advantage, saying that such discoveries and inventions belonged to humanity and should not be in any way hampered by licences and patents. The story of Roentgen's X-rays was carried in morse code signals along telegraph lines and cables to newspaper offices all over the world. In those days, electronic transmission of world news was limited to a global network of undersea cables.

Roentgen made his discovery only a short time after he had decided to study fluorescence in a cathode-ray tube — that is, an evacuated glass tube in which a high voltage is applied between two electrodes. He used black card to mask the tube's luminous glow. Despite this he noticed that, as the tube was switched on, crystals on a nearby table glowed. The tube, he concluded, must be emitting some kind of invisible penetrating rays.

'I have discovered something interesting but I do not know whether my observations are correct', he told a friend. For the next seven weeks Roentgen studied the invisible rays before telling the world of his discovery. He went on to painstakingly investigate the properties of the radiation. Invisible though they were, he was able to describe their electromagnetic behaviour in meticulous detail. In 1901, at the age of fifty, he received the Nobel Prize for Physics.

IN an address by Dr F. Clendinnen, members of the Medical Society of Victoria were told at their August meeting in 1896 that, 'While Professor Roentgen was the accidental discoverer of this sensational phenomenon, it must be borne in mind that he simply completed the labours of others'.[2]

Of course, whether discovery is by chance or intuitive insight is never easy to say. Creative science is usually a blend of keen observation and a quick imagination. Roentgen's casual observation was not so important in itself: it was making the connection that mattered. An English scientist had also observed a photographic plate darken as he experimented with his tube, but he failed to make the connection with what was happening in the tube.

In almost every corner of the globe primitive X-ray machines were soon acquired, or built by the users themselves. After all, the cathode-ray tube had been around for twenty years. Tubes were to be found in many physics laboratories and not uncommonly in the hands of amateur scientists and even artisans.

Professor Thomas Lyle was one who had experimented with cathode-ray tubes. He read of Roentgen's rays in the *Daily Telegraph* while on holidays. Excited by the discovery, he hastened back to his laboratory at Melbourne University to investigate. To his disappointment he found the tubes on hand were built from the wrong glass and they were not evacuated sufficiently. Lyle was not to be deterred from being in at the start of an exciting new field of work. Like many other early scientists, he built his own apparatus. The first object he photographed was a foot volunteered by a colleague, Professor Orme Masson, who happened to be passing by. Revealing the skeleton proved irresistible to the early experimenters studying X-rays.

Thomas Edison, famous for his many inventions, soon adapted a fluoroscope to create skeleton pictures. The X-rays from a cathode-ray tube, after passing through the body, fell on a fluorescing screen. A dark outline of the bones and organs appeared. Edison, always the entrepreneur, put his fluoroscope on show at an electrical exhibition in New York. As people saw their body innards come up on the screen they flinched. Some worried about losing the privacy of their clothes. 'Some crossed themselves devoutly', reported an observer, 'but the great majority came out all smiles and laughter'.[3] Tragically, some practitioners were soon to learn that X-rays were not to be treated so lightly.

The first demonstration of X-rays at a Melbourne hospital took place in June 1896. Dr Clendinnen soon afterwards became the hospital's first honorary skiagraphist, the name then given to an X-ray specialist. He was one of a number of Australian X-ray pioneers whose life was shortened by over-exposure to the rays.

Most early Australian radiologists had a flair for putting together and maintaining 'Heath Robinson' electrical equipment. Technicians who could blow glass, and operate and maintain vacuum pumps and electrical discharge equipment, performed as radiographers. Perhaps because of the isolated and fragmented nature of the small Australian scientific and medical communities, and because European supplies of X-ray machines took months to ship, this technical skill was more necessary in Australia.[4] The early spirit of improvisation was evident at the Sydney Hospital, where Dr Bowker 'sent for Mr Schmidlin, an electrician of Elizabeth St., who kindly came at once and proceeded to take radiographs with the patient on the operating table'.[5] Dr Bowker then successfully extracted a bullet from the thigh of the patient.

To check the image on their fluorescent screen, operators would expose a hand to the X-ray beam. This practice led to lesions on the hand; years later a hand or even an arm would have to be amputated as a result of the lesions turning malignant. Thus, very early in the use of X-rays the evidence was clear that exposure to X-rays could cause cancer.

In 1921 the British Roentgen Society recommended the first safety procedures. The first radiation protection committee in the United States was formed in 1920. But, as no instrument was available to measure the actual power of X-ray beams, the protection that could be afforded was limited. Safety standards were for a time based on a measure of the redness developed on an operator's skin after a certain time of exposure to the beam. From what is now known, operators of those days received very high radiation doses.[6]

By the late 1920s, instruments were available to record the patient's dose, making unnecessary the habit of practitioners exposing themselves repeatedly. A measured dose of X-rays could be administered to a patient. The scatter of rays from the beam could also be detected. Unfortunately, by then many practitioners had become unwitting martyrs to their enthusiasm for X-rays.

NEW gadgets have a fascination for us all. Unfortunately, in our technological age we have the power to do enormously more harm to ourselves and the environment than ever before. This dark side of technology, involving many personal tragedies, has been largely lost sight of in the rush to over-indulge in its benefits — or even to experience it for novelty's sake. The X-ray machine came at the beginning of this age. The abandon with which it has been used has had painful consequences for patient and practitioner alike.

One early victim was the female secretary of a doctor. To make her more attractive, the doctor, in his enthusiasm for his newly-acquired machine, used it to remove hairs from her arms. This led to ulceration and finally to the tragic need to amputate both her arms.[7]

A victim of a macabre piece of quackery submitted to having an X-ray tube thrust against a tumour for the fanciful reason that it 'might have the full benefit of the glow and of its healing virtue'. Within months the patient died in excruciating pain, of a sarcoma caused by excessive doses of X-rays. 'I don't think we know enough of the influences, for good or evil, of the mysterious rays', wrote Dr Thompson, to whom the man had turned for relief. 'It behoves us to deal with X-rays in the most cautious and careful manner.'[8] Unfortunately, such early appeals for caution were lost in an enthusiastic rush to use the new technology.

In the early days of X-rays a diagnostic picture of the spine took more than an hour. A radiologist might break the tedium of waiting by going out to lunch. During these protracted exposures the scatter of X-rays could even irradiate occupants of adjacent rooms.

Hand lesions made operators painfully aware of the serious consequences of large doses of X-rays, called *acute* exposure. But, while the dangers of acute exposure were realised, the much more subtle internal damage to body organs from a succession of small X-ray doses — *chronic* exposure — was learned only decades later from statistical surveys. Many early radiologists received cumulative radiation doses of the same order as people in Hiroshima during the atomic bombing who suffered cancer in later years. Radiologists, like many of their patients, became human guinea pigs, posthumously providing statistics on the health effects of X-rays.

From a study of the obituaries of radiologists, Herman March reported in 1950 that radiologists were at least nine times more likely to die of leukaemia than other medical personnel.[9] A review of the deaths of over 80 000 medical practitioners to the end of 1954 showed that radiologists lived five years less than other medical professionals.[10] A survey of radiologists' children revealed a high incidence of genetic damage.[11]

By the 1930s the worst excesses involving acute exposures had stopped. Even so, until the 1950s health-conscious parents often took the advice of their doctor to have an X-ray check-up on their infants. And still in the 1990s the overuse of X-rays is not just

Table 1.1: Lifetime risk-estimates of some common X-ray examinations

X-ray examination	Age	Male	Female
		One chance in:	
Chest (2 films)	Newborn	3 500	1 800
	10	6 000	4 200
	40	59 000	28 000
	50	1 400 000	630 000
Lower arm (2 shots)	5	300 000	350 000
Angiocardiography (40 films + 30 min fluoroscopy)	5	120	80
	40	800	500
Full mouth (16 shots)	5	600	1 300
	10	600	1 400
	40	2 400	6 100
Full spine — chiropractic (1 shot)	5	570	360
	10	500	300
	40	2 500	1 600
Mammography (2 shots)	20		550
	30		600
	40		1 400
	50		32 200

Source: J. W. Gofman and E. O'Connor, *X-rays: Health Effects of Common Exams*, 1985.

something that happened in the past. The early lessons about the chronic effects of low X-ray doses used in medical and dental diagnoses have yet to be taken seriously by the doctors and dentists who recommend and administer them. The young are the most susceptible: a newborn baby is hundreds of times more likely to contract cancer from a chest X-ray than is a 50-year-old adult. The risk to a newborn child is about twice as great as to a 10-year-old (see table 1.1). The young are the ones who have most dental X-ray check-ups. Full-mouth exposures carry a risk of cancer in later life.[12]

A review of X-ray practices in Victoria in 1982 reported that it was common practice for a lay veterinary assistant to be exposed to X-rays while controlling animals, and that 'the assistant is likely to repeat the procedure on many occasions'.[13] It is not unknown for nurses to hold restless children while they are being X-rayed. Untrained assistants operate X-ray machines. 'They do it behind a lead screen, but the patient is just sitting there.'[14] These practices demonstrate how a cavalier attitude to X-rays persists among medical practitioners. Their credo seems to be: 'Do not do anything too obviously risky, but a little radiation does no one any harm.' These practices are being condoned by public health officials, who need to do much more to encourage practitioners and public alike to keep exposures to the barest minimum necessary to benefit the patient.

Radiotherapy

'It has occurred to me,' a correspondent to an Australasian medical journal wrote in 1896, 'that the physico-chemical action of Roentgen rays might be utilised in hydatid disease (cyst), not as a diagnostic agent, but as a direct *therapeutic* measure.'[15]

Radiotherapy has been tried over the years in a carefree manner for almost every ailment afflicting humankind, including inflamed tonsils and adenoids, ringworm, asthma, tuberculosis of the bone, whooping cough, various skin disorders, arthritis and benign tumours, as well as all kinds of cancer.

One reason for this enthusiasm in the early days was that X-rays were spectacularly successful in curing skin growths not amenable to the limited surgical techniques practised early in the century. This led to ill-judged use of X-rays for internal therapies with primitive equipment and little knowledge of the harm from the radiation. It was a common practice up to the 1940s to use X-rays to treat children supposed to suffer from an enlarged thymus, a gland located in the chest. As a result the thyroid was exposed. Years later a number of these children suffered from benign and malignant thyroid cancers.

The use of X-rays to destroy cancerous or other diseased cells depends on these cells being more readily destroyed than healthy cells. The difference in the response to radiation of healthy and diseased cells can be quite narrow and at times does not exist at all. The more alike the cells, the more similar is the effect of X-rays. This makes it crucial for the X-rays to be beamed accurately onto the diseased tissue. A competent operator, with a modern well-tuned machine (unfortunately not to be taken for granted, even today) will still irradiate some healthy tissue because of internal scatter from the main beam. Some body organs are more prone to generate cancer than others. Reproductive organs are highly radiation-sensitive, but so also are the intestines, bladder, rectum, thyroid, pancreas, and lungs.

Herschel Harris, the chief radiologist at Sydney Hospital, warned in 1914 that, 'X-rays must be regarded as a double-edged weapon. We must learn not to inflict direct injury by the instrument itself'.[16] While X-rays kill cancerous cells, they also cause cancer. A few years after his warning, Harris fell victim to radiation exposure, cutting short his life at the age of 49.

X-rays were used in the 1950s to treat ringworm of the scalp in child immigrants to Israel who came from the poor areas of Europe and the Middle East. The dose of X-rays was large enough to cause epilation (loss of hair). In this way the ringworm was exposed for further treatment. A group of 10 000, who had gone through this treatment as children, were surveyed years later. Though the children lost hair temporarily as a result of the treatment, the X-rays had been 'soft' — they were *low-energy* X-rays — and so had mostly been absorbed in the scalp. Nonetheless the children suffered a high incidence of brain and thyroid cancer in later life.[17]

No dose without some risk

If you question the benefit of an X-ray examination prescribed by your doctor, dentist or chiropractor you are likely to be told that there is nothing to worry about, because the radiation dose is so small. You may even feel foolish to have questioned it. In fact, there will be patients who suffer in later life from X-rays, however small the dose might have been.

In 1955 David Hewitt, a statistician at Oxford University, observed a 50 per cent rise in the number of children in Britain dying of leukaemia in the preceding few years. To find an explanation, Alice Stewart of the university's department of preventive medicine began a survey. She surveyed the mothers of 1 694 children who had died of leukaemia or cancer in the previous two years. An equal number of mothers with healthy children were used as a control group.

It emerged from the survey that the major difference between the two groups was in the number of X-ray examinations each had experienced. Stewart and her co-workers found that twice as many children who were born of mothers given pelvic X-rays died of a malignancy before the age of ten.[18] Yet the mothers whose children had died of cancer had been given quite low doses of X-rays. Up until the survey it was assumed in medical circles that X-ray doses below a certain threshold were harmless, though there was no scientific evidence to back up that assumption. In fact, there was already evidence to the contrary. Now the work of Stewart provided convincing evidence that, however low the dose, there was a risk of inducing cancer.

The evidence for harm was presented carefully. Yet the immediate reaction within the medical profession was disbelief and rejection. The finding that extremely low doses of radiation to a foetus could produce a malignancy, far from being immediately acted on to protect unborn children, became instead the target of acrimonious criticism. Doctors did not welcome questioning of their habit of ordering X-rays for multiple and difficult births, and for even less substantial reasons during pregnancy.

For a time Alice Stewart's survey funds were curtailed. Despite the opposition, Stewart persisted with her survey. In 1970 she again published results, this time showing that irradiating a foetus in its first three months increased the chance of developing cancer by ten times. Even a single X-ray increased the chance of a child malignancy.[19]

Meanwhile, a survey published in 1962 by Dr Brian McMahon

at the US Harvard School of Public Health had compared the children of 70 000 mothers who had been X-rayed during pregnancy with children of mothers who had not been X-rayed. McMahon found that the incidence of cancer was 40 per cent higher among children of mothers who had been X-rayed during pregnancy.[20] In 1985 a study in the United States of twins born between 1930 and 1969 again confirmed the increased chance that children whose mothers had been X-rayed would suffer from leukaemia and other childhood cancers.[21]

The finding of Stewart that unborn children exposed to X-rays have an increased risk of developing childhood cancer is today gospel. But Stewart and other scientists had to battle for many years before the medical profession acknowledged the dangers of giving pelvic X-rays to pregnant women.

Even so, the International Commission on Radiological Protection (ICRP), which recommends on medical radiation practices, has recently relaxed an important rule introduced in the 1970s regarding the irradiation of women. This rule is to limit X-ray diagnoses of women to the ten days after the start of their period to avoid the possibility of irradiating a very young foetus. In 1983 this rule was repealed by the ICRP without explanation. Since then the ICRP has ignored the case put by a large number of scientists for restoration of the rule.[22]

Ultrasound imaging of the foetus — a technique which uses high-frequency sound waves that are not part of the electromagnetic spectrum — is now mostly used instead of X-rays. This is done largely as a matter of routine. Adverse birth effects have not been observed to result from this technique; but there has not been any extended health study of the practice. From the few animal studies carried out, ultrasound has been shown to reduce the weight of the foetus and to alter the behaviour of the offspring.[23] So far, the only indication of an effect on humans has been from a preliminary study in the United States, which indicated a possible reduction in foetal weight.[24]

The routine use of ultrasound on the foetus reflects how new technologies are taken so much for granted. That is not to say that ultrasound is not almost certainly safer for internal imaging than X-rays. However, until the biological effects of ultrasound are assessed fully, would it not be prudent to use it only when symptoms indicate a need for internal imaging?

X-RAYS create electrically charged atoms — that is, *ions* — in the matter they penetrate. They share this *ionising* ability with the radiations emitted by radioactive materials.

It was the late 1970s before national and international bodies responsible for setting standards for radiation safety belatedly adopted the concept that all ionising radiation exposure carries some risk of cancer, and that there is no such thing as a safe dose of ionising radiation.

In 1927, Professor Muller began his now famous work on the genetic effects of X-rays on *Drosophila* (fruit fly). He observed severe genetic deformities at high radiation doses. At lower doses he observed lesser changes in form and appearance which still left the fly able to grow and develop in the normal way. These changed characteristics are the result of *mild mutations*. The possibility therefore exists that mild mutations will accumulate from generation to generation, raising serious questions about the long-term genetic deterioration of exposed species, including human beings. Therefore, any additional radiation exposure of the human population must necessarily be viewed with deep concern.

Ever since the earliest genetic work, the idea that mutation was proportional down to the lowest radiation dose has been entrenched in genetic thinking. It soon became evident to geneticists that a concept of a safe dose of an ionising radiation did not make sense. But people in the business of setting acceptable limits of exposure have found the 'no safe radiation dose' concept inconvenient. Even someone like Professor Edward Radford, who became something of an outcast from his scientific community because he consistently argued that radiation hazards were being underestimated, confesses that he too has some difficulty with the concept.

Radford actually played a key part in having the concept of 'no safe radiation dose' grudgingly acknowledged by US scientific bodies setting radiation safety standards. He admits to a basic contradiction between the idea of radiation safety standards and the knowledge that any exposure to an ionising radiation, including X-rays, will cause genetic damage to offspring in future generations.

'It has been very difficult, and I include myself in this,' Professor Radford told the Royal Commission into British Nuclear Tests in Australia, 'to incorporate into a radiation protection standard the limits that must be set in order to prevent damage to the next generation, or five generations hence, or ten generations hence. It quickly gets out of our realm of experience where we simply brush it aside.'[25]

Radford's prick of conscience relates to a most challenging issue of our times: the preservation of the human genetic code. It is one facet of the wider struggle to hold the line against further deterioration of the genetic bank for the whole biosphere, on which the global ecology depends for its healthy reproduction.

The nuclear connection

In the mid-1970s, Dr Rosalie Bertell, a nun and a mathematician, agreed to speak at a public meeting about her researches. She had taken part in the Tri-State Leukaemia Survey, one of the largest epidemiological studies on the connection between X-ray exposure and disease. She had estimated from the survey data that individuals 'with asthmas, severe allergies, heart disease, and diabetes were about 12 times as susceptible to radiation-related leukaemia as were healthy adults'.[26]

The finding raised the issue that some sections of a population are more vulnerable to radiation than others. Ailing children especially could suffer more than the average person. Of course, what Bertell had to say about medical radiation practices also applied to the emissions from the nuclear industry's radioactive wastes.

The meeting had been called to oppose the building of a nuclear reactor at Buffalo in the United States. Nervous about being on a public platform for the first time, Bertell simply spoke about her findings in a modest and factual way. Thereupon she became the target of retaliation by the nuclear industry. 'It was a surprise,' she recalled, 'mail opened, a nasty article in the local paper, a fuss at the institute where I worked, led by a man whose research was financed by the defence department'.[27]

Two other researchers involved in the survey, Drs Bross and Natarajan, also pointed to likely long-term genetic effects of low-level radiation on the human species.[28] As a result, their funds were withdrawn.

Stewart, Bertell and other medical researchers found that the nuclear industry was a much more powerful and punitive opponent than entrenched medical interests. At the time, nuclear officials were playing down the health damage from the 1950s nuclear weapons testing. They were promoting nuclear power as clean and safe. Bertell's findings could be the undoing of their public relations campaign.

Research findings adverse to the use of radiation technologies have often been kept classified as secret, perhaps for decades. Funding authorities, often a defence agency, can suddenly say, without explanation, that further research into a particular field is no longer needed. What people do not know they cannot worry about.

In 1979 Irwin Bross, a victim of research funding manipulation, told a US Congressional hearing into the effect of radiation on human health that some 'scientists, physicians, weapons technologists and engineers, whose livelihood depends on radiation technology' were guilty of active or passive complicity in the suppression of unwelcome findings. In some instances, he said, the bias has influenced scientific journals. The result has been a 'successful suppression of the truth about low-level ionising radiation for 20 years or more'.[29]

Unproductive X-rays

In 1984, at the outpatients' department of a major Melbourne hospital, a doctor was presented with a patient — me — who had jabbed a small piece of lead from a pencil into his finger. 'Let's have an X-ray', the doctor suggested. I objected that an X-ray was hardly needed for such a simple matter. The doctor still felt that it would be advisable, 'just to be sure we can see everything'. After some debate the lead fragment was extracted, without X-rays, by probing with a simple instrument. This was a trivial incident, but it shows that the institutional habit of overusing X-rays is still current.

X-ray machines were used until the 1960s for shoe fittings.

Public authorities stopped the practice only after a campaign by concerned scientists. The use of X-rays to detect the lung disease tuberculosis was once compulsory in Australia. The mobile X-ray

UNPRODUCTIVE X-RAY EXAMINATIONS

The chest X-ray is the 'most frequently performed radiological procedure throughout the world, accounting for up to 50 per cent of all diagnostic radiology.' Chest X-rays are used routinely to detect cardiovascular diseases and lung cancer on admission to hospitals before operations. They are also used on certain occupational groups.

Chest X-ray during pregnancy, 'however well conducted, provides some radiation exposure to the embryo or foetus'. X-ray exams are also used frequently for following symptoms of minor head injuries such as headaches and low back pain or to investigate 'vague abdominal discomfort'.

Studies have shown that, in *all* cases where an X-ray exam is given to an asymptomatic person (without symptoms), or to a sick person whose illness can be diagnosed in other ways, the exam has proved to be 'clinically unproductive'.

The following are guidelines on the limits of usefulness for chest, skeleton and abdominal X-ray exams.

Chest

Occupational groups: Except where there is an epidemic in a country, 'the routine examination of educators, students, health service personnel, food handlers and others without specific occupational respiratory hazards is very unproductive'.
Preoperative: Studies show that the results of chest X-rays 'did not stop the operation, did not alter the type of anaesthesia and did not alter the post-operative complication rate'.
Cardiovascular diseases: 'In the investigation of patients with cardiovascular diseases — in particular those with hypertension, angina and myocardial infarction — clinical examination is more useful than chest radiographic examination.'
Lung cancer: 'There is no evidence that periodic chest radiographs in symptom-free individuals produce significant benefits or aid in the early detection of primary cancer of the lung.'
Acute pneumonia: 'Repeat chest radiographs are only indicated if the patient does not progress satisfactorily.'

machines used were likely to have been poorly aligned because of jolting during transport. The radiographers carrying out the millions of diagnoses would have been constantly exposed to

Paediatrics: 'The mere fact that a child is sick is no indication for a chest radiograph in the absence of cardiopulmonary abnormalities.'

Mammography: 'Periodic mammography of asymptomatic women should be restricted to those aged 50 years or over. Under the age of 50 years periodic screening is not advised unless there is a personal or family history of breast cancer or other high-risk indications.'

Skeleton

Head injury: 'Skull radiography is not recommended in patients with mild head trauma who are asymptomatic or who present with one or more of the following symptoms: headache, dizziness, simple scalp laceration, haematoma and contusion or abrasion.'

Back pain: 'The examination in lumbosacral pain is of no medical benefit to the individual. The weight of evidence indicates that radiological findings are of no predictive value for the occurrence of future disability and have no effect on the subsequent incidence of low back pain.'

'Pre-employment lumbosacral radiography should not be done.'

Abdominal

Accident: 'There is evidence of overuse of films taken in the supine and erect positions which are often treated as a routine combination.'

Gastrointestinal tract: 'There is evidence of overuse of barium examinations . . . all barium studies should be carried out by qualified radiologists using appropriate fluoroscopic equipment.'

Intravenous urography: 'There are circumstances in which intravenous urography is not clinically useful and should be strongly discouraged.'

Source: *A Rational Approach to Radiodiagnostic Investigations*, a World Health Organisation publication, Technical Report 689, 1983.

scattered X-rays. Public health authorities continued to promote the X-ray examinations after the disease had been virtually eradicated and the exercise no longer made sense. When no longer compulsory, state health departments turned for a time to advertising 'free' chest X-rays.

Today some institutions still make an X-ray examination an obligatory part of routine health checks. Patients admitted to public hospitals are routinely given an X-ray. Readmitted only a few months later, they will be X-rayed once again. Pathology clinics are known to be advising women of all ages to have a mammogram each year, despite the advice of the World Health Organisation that 'periodic examination of asymptomatic women should be restricted to those 50 years or over' (see box: Unproductive X-ray examinations).

Persistent overuse of X-rays is one cause for concern; the appalling overexposure of patients during X-ray examinations is another. A fairly general carelessness in operation and maintenance of X-ray machines has been revealed by a survey of radiation practice in 300 medical institutions. The survey showed that the radiation dose for a chest X-ray varied by a factor of up to a thousand between the minimum and maximum dose. It was estimated that more efficient operation of X-ray machines could reduce the collective exposure of the Australian population to X-rays by 40 per cent. This 'will reduce the number of possible radiation-induced cases of cancer and genetic defects in the population'.[30] If the overuse of X-rays is also taken into account it seems that the collective exposure could be reduced to a very low level indeed.

Even so, the survey covered only clinics where radiologists were in charge. What of medical, dental, chiropractic and other professionals who have had minimal training in the operation of their machines? What of the continued use of poorly focussed portable equipment? Unlike all other kinds of instruments affecting our well-being, governments have been tardy in regulating X-ray machines for accuracy and safety.

By the mid-1980s most state authorities had passed legislation to impose stricter regulation on radiation sources, including X-ray machines.[31] It remains to be seen what will be the effect of new regulations. The medical profession is unwilling to register the radiation dose received by patients. In fact the ICRP actually opposes routine reporting of radiation dose in diagnostic practice. Until the profession is willing, Australians' collective X-ray dose cannot be known. Yet this information would be helpful in the fight to limit the incidence of cancer and genetic disease.

The only indicator of the total population exposure is the steep increase in consumption of X-ray film. Between 1950 and 1970 Australian consumption of film went from 60 000 to 2 100 000 square metres — an increase of over 3 500 per cent, while the population increased only 52 per cent.[32] While a taboo exists on registering the patient dose it is unlikely that the cavalier use of X-rays by individual practitioners will be changed much by legislation.

THE overuse of X-ray technology is now firmly entrenched in modern medical practice. According to a 1983 World Health Organisation (WHO) report, *A Rational Approach to Radiodiagnostic Investigation*, 'patients have come to believe that no examination by their doctor is complete unless they have been X-rayed. The procedure is satisfying because it is usually dramatic and causes little discomfort or inconvenience'. For the doctor, 'requesting an X-ray has become a comforting ritual'.[33] According to the WHO report, not enough is being done to critically appraise what benefits are being gained from the greatly expanded usage of diagnostic X-rays over recent years.

A study in 1987 in the United States of 7 000 patients with head injuries showed that, while 58 per cent had been given a skull X-ray, only two per cent of adults and four per cent of the children were found to have skull fractures. The absence of fracture had been clinically obvious in a high proportion of the X-rayed patients. Similarly, in Australian hospital casualty-wards, doctors routinely order skull X-rays for accident victims. This can be against the advice of hospital X-ray audit committees. The doctors' defence is that they fear legal action if they happen to miss diagnosing a fractured skull.[34] This is one disadvantage of the growing trend towards medical litigation.

X-ray and other radiation-based diagnoses over the years have come to rely on ever more complex and expensive equipment pushed by entrepreneurs in medical technology. The rapid proliferation of CT scanners has caused a dramatic rise in the collective radiation dose.[35] Medical institutions feel compelled to acquire the latest technology; then practitioners become so captivated by its power to produce images that they overuse it uncritically. While individual patients have undoubtedly benefited from new diagnostic techniques, it is also true, as the WHO report goes on to point out, that the added costs have become a great burden on national health services. This has drained basic health care of financial support.

To accept or reject a doctor's advice to have an X-ray examination can be a difficult decision. It is helpful to remember that a doctor's advice is as much a matter of personal judgement as expertise. Many routine X-ray examinations have been demonstrated to be clinically unproductive (see box: Unproductive X-ray examinations). Quite often the medical indications are available from other kinds of investigations. If an X-ray is clinically indicated, there is now a great deal of information available on how to minimise exposures while obtaining optimum results.[36]

Parents have a particular responsibility when deciding about an X-ray for their children. It is the young who are most vulnerable to radiation, though the consequences may show up only in later life. In general, parents agree to too many full-mouth X-rays for their children, thus increasing the chance of thyroid cancer.[37]

None of this is to say that X-rays, since discovered by Professor Roentgen, have not brought immeasurable benefit in cure of injury and disease. They have. X-rays now make an indispensable contribution to health care. But a patient's decision to have an X-ray examination should be an informed one.

On a global scale, according to the WHO report, X-ray diagnosis 'is the largest manmade source of population exposure to radiation. In the world as a whole it contributes approximately 20% of the natural background radiation and in developed countries this figure approaches 50%. No one concerned with this problem can afford to be complacent'.[38]

Back in 1914, Sydney Hospital's chief radiologist, Herschel Harris, warned that X-rays could be a double-edged sword. He was right.

Chapter 2

Our new electromagnetic environment

> By the beginning of the last half of the twentieth century man-made electromagnetic frequencies were the overwhelmingly dominant constituent of the earth's electromagnetic environment. With the benefit of hindsight we can now see that it was dangerous to have made such a drastic alteration in our environment without first studying the potential biological impact.
>
> — *R. O. Becker & A. A. Marino,* Electromagnetism & Life, *p. 177, 1982.*

BY early this century, X-ray diagnosis and radiotherapy were well-established medical practices. Only eighteen years after the discovery of X-rays, Sydney Hospital's chief radiographer, Herschel Harris, told the 1914 Australasian Medical Congress that 'well over 30 000 cases have passed through my hands treated with the X-rays and radium'.[1]

Meanwhile, the invisible radio waves discovered in 1888 by a German physicist, Heinrich Hertz, were being tested to carry signals in the way semaphore uses pulsed (visible) light.

Hertz's discovery of radio waves was not by chance. Twenty years earlier, a Scottish physicist, James Clerk Maxwell, theorised that visible light travelled as waves. He called the wave system electromagnetic radiation (EMR) because it was propagated by a coupling of *electric* and *magnetic fields*. The waves can be imagined as speeding through space by the two fields leapfrogging over each other. Maxwell predicted that other EMR, invisible to the human eye, would travel at the same speed as light, but at varying wavelengths.

Hertz made the first electrical transmitter in which sparks jumped between two electrodes, thus causing electromagnetic waves to be broadcast. The invisible waves were picked up by a receiver located at the other end of a large room. The wave energy induced an electrical current to flow in the receiver, generating sparks across its two electrodes. Hertz actually measured the length of his invisible waves as 660 millimetres (see box: Counting electromagnetic waves). Waves of this length are in the *radio frequency* (RF) band and today are used for broadcasting television (see box: The electromagnetic spectrum).

Included in the RF band are radio waves and microwaves. Microwaves are used for satellite communication, radar, and domestic ovens.

Whereas cathode-ray tubes had been around for many years ready to be adapted as X-ray machines, radio broadcasting took thirty years to develop. For, while Maxwell's electromagnetism laid the theoretical foundations of modern telecommunications, much had to be learned about high-voltage electrical sources and radio valves, tuning coils and other electronic devices, before it was practical to transmit over long distances and amplify the broadcast signals.

IN 1895, a young Italian inventor, Guglielmo Marconi, transmitted radio waves between points about a kilometre apart. Within two years, he had enlisted the support of the British Post Office and, by 1898, he could send messages across the English channel. In 1901 he transmitted a message across the Atlantic Ocean. The radio wave carried the single morse code letter "S". Marconi used the spark generator to transmit *pulsed* wave signals. In 1909, Marconi received the Nobel Prize for Physics.

The first overseas transmission from Australia was sent in 1906 from Point Lonsdale on the southern coast. The equipment was supplied and operated by the Marconi Telegraph Company, founded by Marconi in 1897.

It was not until an American physicist and engineer, Lee De Forest, invented the triode radio valve in 1907 that it was possible to broadcast *continuous* waves. These waves act as carriers that

COUNTING ELECTROMAGNETIC WAVES

Electromagnetic radiation (EMR) can be defined according to either its *wavelength* or its *frequency*.

The *wavelength* is the distance between the crest of one wave and the crest of the next.

The *frequency* of the wave is the number of waves which pass a given point each second. If one crest of a wave passes point A (in the diagram) each second, then the frequency of the wave is said to be one cycle a second. The unit of one cycle per second is called a Hertz (Hz), named after Heinrich Hertz, the discoverer of radio waves. If 10 wave crests pass each second then the frequency is 10 Hz.

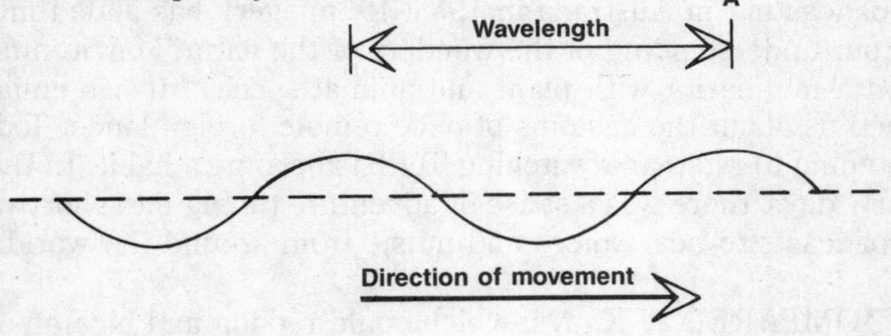

The wavelength multiplied by the frequency equals the speed of the radiation. Since the speed of EMR is constant at 300 000 kilometres a second we can calculate either wavelength or frequency of a particular radiation if we know the other one.

Wavelength × frequency = 300 000 km/second
Frequency = 300 000 ÷ wavelength
Wavelength = 300 000 ÷ frequency

If a group of people all walk at the same speed of 4 000 metres each hour, but with longer or shorter steps, we can calculate the different *stepping frequency* for each person. Thus a person walking with a step of 1 metre (equivalent to wavelength) takes 4 000 steps/hour (frequency). Another person walking with a step of 0.8 metre takes 4 000 ÷ 0.8 or 5 000 steps/hour.

In the same way, if a radio wave has a wavelength of 2 kilometres then its frequency is 300 000 ÷ 2 or 150 000 Hz. That is, 150 000 wave crests pass point A each second.

'piggyback' signals conveying the human voice and music. In 1916 De Forest was responsible for transmitting the first radio program.

By the early 1920s people were 'tuning-in' in their own homes. 'Listening-in' to music and voice issuing forth from the megaphones of their wireless sets was a thrill older people still remember. How crackly with static the reception could be! Radio has brought people together from all parts of the globe without the need for an almost endless stretch of wires. Sound broadcasting over radio waves has enabled an immediate awareness of current events.

Television services using ultra-high-frequency waves began broadcasting in Australia in 1956. Its imagery has added much to our understanding of the wonders of the natural environment and a familiarity with plant and animal species. It has enlightened us about the customs of once remote foreign lands. Today, listening to radio and watching TV has become a habit. In those early days, there was a sense of adventure tuning into shortwave broadcasts to hear voices and music from around the world.

COMPARED to X-rays, which could redden and ulcerate the skin, the EMR used in early broadcasting must have seemed quite innocuous. No one could sense radio waves, and the pioneers of broadcasting would have given little thought to the likelihood of health damage from their transmissions.

However, transmitters broadcast out to the wide environment. Their radiation pervades the space we enter or even inhabit for most of our lives. Marconi's signal across the Atlantic not only gave birth to modern telecommunications, it heralded a marked change in the earth's electromagnetic environment. Another source of change was the electricity distribution systems spreading across the land to supply proliferating domestic and industrial electrical equipment. Radiating out from all electrical systems are invisible electric and magnetic fields, which oscillate back and forth in time with the alternating current. We are all — city dwellers more than others — immersed in these invisible fields.

As transmitters became more powerful it became evident that exposure to fairly intense beams of radiation, especially microwaves, could cause injury by heating body tissues. But what of the EMR pervading the environment and our bodies? Though it is too weak to cause acute *thermal* injury, the question began to be asked: was it so innocuous after all? The first suggestion of biological changes more subtle than simply overheating tissue came not from the experts, but from those closest to potential health problems: those on the job.

During World War II it was lore among sailors that radar — which uses the microwave band — caused sterility. Radar operators would charge a fee for a dose of 'contraceptive' radar administered to sailors going on shore leave. A group of workers exposed to radar believed that the balance between the two sexes among their offspring had shifted from parity to a predominance of girls. It is now known that microwaves can affect sex organs.

When some people said that they had an exceptional ability to hear radar it was dismissed as nonsense. A scientist who agreed with them was dismissed by other scientists as a crank. It is now known that some people can hear radar at certain frequencies; all that is questioned is how they hear it. Radar operators, during and after the war, complained of headaches, dizziness, eye-strain and insomnia. In the early 1950s a medical report to an aircraft corporation described cataracts, headaches, brain tumours and heart conditions as possible ill effects from exposure to radar. A report of a study in 1954 of 200 radar workers indicated blood anomalies. Later this finding was attributed to 'a variation in interpretation by a laboratory technician'.[2] Generally complaints of this kind were dismissed by health and safety bodies as too subjective to warrant further investigation.

In the Soviet Union similar symptoms among people working with radar and with electricity systems were taken more seriously. The result has been that the Soviet Union applies much stricter limits to the exposure of workers to radio waves and microwaves in the RF band than does the West. Western scientists berated Soviet scientists, as well as the few in their own ranks, who took seriously health effects of low-level exposures to RF waves.

In the last few years it has been found that RF waves can influence endocrinal hormone secretions. This could possibly explain at least some of the 'subjective' symptoms complained about by radar and electrical workers forty years ago.[3]

Contention, and often acrimonious debate over the health effects, has always been in the background of the setting of health and safety standards for RF exposure. Standards are set by expert committees after considering the data from research in the field. Where there is conflicting evidence the permissible exposures finally arrived at depend on which research findings are accepted and which are dismissed. These considerations are greatly influenced by personal judgements of what constitutes 'objective' research and whether borderline evidence is taken into account. If borderline evidence of adverse health effects is rejected, and then confirmed by later research, people will have suffered because of inadequate protection from safety standards.

IF our eyes were able to sense the waves of invisible radiation now emanating from the innumerable communications and other electronic sources, we would be immersed in a shimmering haze. Shafts of more intense light would streak like searchlights across the scene from radio, TV and radar towers. If these phantom lights left their marks as colour stains we would, no doubt, have been more cautious about intruding them into the environment and our lives.

Some individuals claim that they can feel the electric field under high-voltage power lines like the sensation of a breeze blowing on the skin. Most of us, however, sense nothing of the EMR environment in which we live. We can get some feel for its pervasiveness by the way we tune into broadcast programs wherever we carry our radio. Our own bodies either pick up or are transparent to EMR. Up to a hundred years ago humans lived in a quite different radiation environment.

EMR comes to us from all directions. Our short-wave programs may come from the other side of the globe, bouncing back and forth between earth and ionosphere, a mantle of positively charged particles. Our television signals are carried on line-of-sight microwave beams from antennas poised on hills and mountain tops. Radiotelephony between cities is relayed on microwaves between dishes on hilltop towers. Communication satellites in space are linked to their earth stations by microwaves. Proliferating cellular telephone systems cast a veritable haze of radio waves over their own cities. Military forces keep 'enemy' defence systems under constant surveillance using powerful radar. Airports control aircraft by tracking with radar. Microwave beams from radar transmitters reflect off aeroplanes back to their source because, unlike light beams, they are not dispersed by dust in the air. However, some microwaves can spill from the dish.

Microwaves are used in the cooking of food and as a source of industrial heat. Leaking industrial equipment has been known to make the workplace environment alive with electromagnetic fields. Offices and news rooms have been turned into factory-like production lines with their rows of visual display terminals (VDTs). This book, for instance, was written and edited on a word processor. Like all electronic equipment, VDTs add their contribution to the radiation environment. Have we adopted so much electronic gadgetry from genuine need or because we have been induced to feel it is indispensable?

When we switch on the electric light or warm ourselves by a radiant heater we sense the visible and infrared radiations. However, most of us are blissfully unaware of the electromagnetic fields emanating from high-voltage and distribution lines bringing electricity to our houses. Electromagnetic fields envelop the electrical appliances we use. Asleep on a 'live' electric blanket, we bathe in these fields.

LIFE evolved in a vastly different radiation environment to the one we live in today. In past ages the environment was pervaded by very weak EMR in the radio frequency band; it has always been exposed to the earth's geomagnetic field. Over the years the effects on the human body of fluctuations in the strength of the geomagnetic field over the earth's surface have been investigated. It is now believed that some birds' pineal glands are sensitive to the field, assisting the birds to navigate over long flights.

Natural radio frequencies are at greatest intensity during thunderstorms. But telecommunications systems now contribute most RF emissions to the environment. The exposure of the human species to RF waves is now millions of times greater than before our radiation age.

We know that certain extra-low-frequency (ELF) electric fields

THE ELECTROMAGNETIC SPECTRUM

When we tune into a radio program by moving the pointer along AM or FM scales of our radio dial we are adjusting a tuner to receive the radio waves or microwaves being broadcast by a particular radio station. The waves we seek to pick up will have a specific *wavelength* and *frequency*. They will be in the *radio frequency* (RF) band of the electromagnetic *spectrum* at the point where the radio station has a licence to broadcast.

The spectrum is divided arbitrarily into blocks of wavelengths, or *bands*. The names along the spectrum are not consistent. Thus the ultra-high-frequency band is at the low-frequency end of the spectrum. We can think of the EMR spectrum as a banded measuring-stick with each band representing a range of wavelengths or frequencies.

At one end of the spectrum lies the *extra-low-frequency* (ELF) band, which is generally taken to cover frequencies below 3 000 cycles a second, or 3 000Hertz (Hz). The alternating current flowing along electric power lines emits ELF at a frequency of 50Hz. Because they have such low frequencies ELF waves stretch great distances. A single ELF wave may stretch thousands of kilometres (see box: Counting electromagnetic waves).

Next to ELF is the RF band. Radio waves in the RF band are allocated to provide for broadcasting services. The 'long-wave' (low frequency) end of this band between 1 500 and 600 metres (2×10^5 to 5×10^5Hz) is used in Australia for aircraft communication. The *very-low frequencies* (VLF) range up to about 100 000 (10^5)Hz. VDTs emit VLF around 14 000Hz.

The familiar radio 'broadcast' band ranges from 600 to 150 metres (5×10^5 to 2×10^6Hz or 2 megaHz). 'Shortwave' broadcasting uses wavelengths ranging from 150 to 10 metres (2 to 30 megaHz). After the shortwave band is the *very-high-frequency* band (VHF), ranging from 10 to 1.5 metres (30 to 200 megaHz); and the *ultra-high-frequency* band (UHF), ranging from 1.5 metres to 300 millimetres (200 to 1 000 megaHz). These two latter bands are used for television and FM broadcasting. (To express extremely large numbers, a superscript indicating the number of noughts is attached to the 10. Thus 1 000 000 is 10^6.)

Then comes the microwave band, whose wavelengths range from less than a metre to a fraction of a millimetre. The microwave frequencies are used for heating, satellite communication, and radar. UHF is at the top end of the frequency range of radio waves, and is often referred to as being at the low end of the microwave frequencies.

Moving further along the spectrum, microwaves merge into the perceptible natural radiations: *infrared* and *visible* light. It is as though there are two very narrow windows in the EMR spectrum. Through one we can feel the warmth of infrared rays. There are cameras which 'see' with reflected infrared light so that they are capable of taking pictures in the dark. Through the second window we sense visible light with the retina of our eye, which gives us our images of the world around us. Our eye is able to break up this very narrow visible band into an infinite variety of colours.

Beyond the visible light band is the *ultraviolet* (UV) band. We are made aware of its presence in sunlight by the way it tans our skin. This band is divided into near-ultraviolet (next to the visible) and far-ultraviolet (away from the visible). Ultraviolet and lower frequency radiations are *non-ionising* EMR.

Next higher in frequency to the UV rays are the penetrating X-rays emitted from cathode-ray tubes that are operated above 15 kilovolts. The higher the voltage the higher the energy of the emitted X-rays. Industrial radiography uses 'hard' X-rays to scan metal castings and welds. Such X-rays are emitted from tubes operated up to 150 kilovolts. Tubes in colour television sets use 40 kilovolts, which is high enough to produce incidental 'soft' X-rays that in well-designed tubes are mostly, but not completely, absorbed.

Gamma rays, which emanate from radioactive substances, are also electromagnetic radiation. The gamma band reaches up to extremely high frequencies (and hence correspondingly extremely short wavelengths). In applications where high energy radiation is required in industrial radiography, either 'hard' X-rays or gamma rays are used. We deal in greater detail with gamma rays when we come to radiations from nuclear sources.

For the purpose of following the radiation controversy it is useful to know the order in which the bands fall along the spectrum: extra-low frequencies, radio frequencies (radio waves and microwaves), infrared, visible light, ultraviolet, X-rays and gamma rays.

exist naturally. They arise from interaction between the earth and the ionosphere. Significantly, they have similar frequencies to those associated with some internal functions of the human body. The body's own electric fields have been measured. Most familiar is the electrocardiograph, which measures heartbeat frequencies between 0.2 and 100Hertz. The electromylograph measures signals from muscle fibres, with frequencies between 100 and 3 000Hz. The electroencephalogram measures brain waves, with frequencies between zero and 50Hz. (see figure 2.1.)

Figure 2.1: The frequencies of some of our body signals occur in the same range as the earth's extra-low frequencies

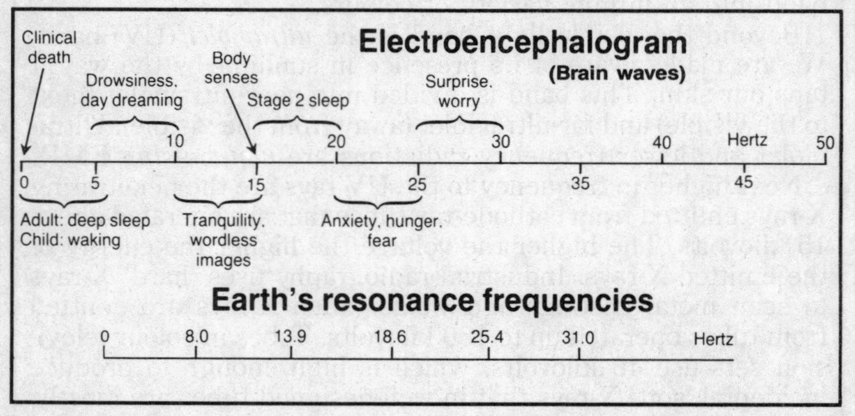

There is a growing body of evidence that living creatures have emerged from their long evolution 'tuned in' to certain naturally occurring pulsing electromagnetic fields.[4] The electromagnetic fields resonate at frequencies in the same range as those observed in the human brain. Body organs are now thought to have evolved with sensitivities and dependencies on specific natural frequencies. One hypothesis is that the evolution of early protein units, before life itself began, emerged as a result of tuning in to these fields on primordial earth. Marine creatures can have an extraordinarily high sensitivity to natural electrical fields, some using them to navigate or detect prey.

One hypothesis is that living cells have the means to receive and amplify the natural signals and use them in communication systems, co-ordinating their metabolic functions. It could also be reasoned in reverse that, because life evolved with only certain frequencies, introduced frequencies could be a source of stress to living creatures. The possibility of stress effects is reinforced by recent findings of laboratory and epidemiological studies on effects of artificial EMR. However, these findings are hotly debated in the scientific community.

Changing environmental values

By 1970, broadcasting and radar transmitters had not only become a familiar sight on the landscape but were often extremely powerful. Questions about the impact of the scatter from their beams on the health of residents living near transmitters could no longer be ignored. People were also becoming more conscious about the way their health was being threatened by pollution.

Possibly the first public alert about RF emissions was in 1973, when the US Consumers Union questioned the safety of microwave ovens.[5] The potential for thermal harm from microwaves is evident from the way they cook food. The Australian Consumers' Association, in its November 1977 issue of *Choice*, reported that one model of microwave oven leaked radiation. Two years later the association said that 'whether the [safety] standard is adequate or not is still a matter of debate'.[6] Once sown, the seeds of doubt about the health effects of microwaves soon sprouted community concern about microwave relay towers and radar transmitters situated in suburban neighbourhoods.

Another influence in the 1970s was the growing interest taken by trade unions in occupational health problems. A rapid transformation in the workplace environment was occurring with new technologies involving toxic chemicals, radioactivity and RF waves.

What had not escaped the notice of some trade union health officials was the vastly lower RF exposures permitted by Soviet radiation safety standards compared to those in the West. However, it seems that Soviet military activities are exempt from the strict national standards. If so, then the setting of civilian standards is freed from the kinds of military influences that operate in the West.

The Soviet attitude to occupational health apparently differs from Western attitudes. It has been said that in the Soviet Union there is much more concern for the physical and mental fitness of workers.[7] In the West, health and safety regulation is restricted very much to the prevention of serious accidents and disease. This difference in attitudes could explain why Soviet health and safety bodies have paid more attention to the chronic ailments complained about by radar workers. However, this understanding of Soviet philosophy on workers' health is contradicted by the way workers were so blatantly exposed to high radiation doses in the clean-up of the crippled Chernobyl nuclear reactor (see chapter 11).

Certainly, Western health-related RF research has proceeded differently from that in the Soviet Union. In the West, scientists have devoted most effort to investigating severe health damage from fairly intense exposure to the microwave part of the RF band. Yet this research will not say anything about the health effects of the continuous low-level exposures, often over a lifetime, experienced in the workplace and the wider environment. Since no one is to be seen actually dying directly as a result of low-level exposure, this area consistently has been given low priority in research funding. Meanwhile, the lack of admissible evidence of health damage from low-level RF emissions meant there was no need for stricter standards.

However, by the 1970s independent researchers were demonstrating that low-level exposures can cause changes in chromosomes, blood cells, hormone secretions, and calcium transport in brain cells. Finding these biological effects gave some support to Soviet claims of adverse health effects, although such claims were still outside mainstream thought on the subject.

In 1977 the American investigative journalist Paul Brodeur wrote his provocative book, *The Zapping of America: Microwaves — Their Deadly Risk and Cover-Up*.[8] The book brought the scientific controversy into the political arena. Community groups and trade unions were alerted by Brodeur's revelations of the suppression of research findings. They began to question the effectiveness of existing safety standards. Brodeur's most startling revelation concerned the Moscow embassy affair.

The Moscow embassy affair

At a superpower summit in June 1967, microwaves were the subject of a confidential exchange between the US President Lyndon Johnson and Soviet Prime Minister Alexei Kosygin. President Johnson asked that the Soviet Union stop irradiating its Moscow embassy with microwaves and harming the health of American citizens. The US defense department first discovered microwaves were being beamed across the street into the embassy in 1962. Despite the request, irradiation of the embassy continued on and off in the 1970s and was resumed for a time in 1983.[9]

The intensity of microwaves reaching the embassy was 500 times less than the level workers could be subjected to in the United States. On the other hand, the exposure was twice the highest limit set in Soviet standards. But if US officials really believed in their own public statements about how safe microwaves were, they could hardly conclude that the Soviet's microwave beams were undermining the health of embassy staff. Yet that is exactly how they did respond. They could not say so publicly, because that would raise embarrassing questions about the safety of microwave practices in the United States. It was one supposed harassment the United States preferred not to use to score points in the superpower propaganda war. The approach of US officials to the microwave affair was to keep the whole matter strictly secret while they investigated.

In 1966 a covert study, called *Project Pandora*, was commenced. The health of embassy staff, who were blissfully unaware of the real reason for a rash of official concern, was kept under regular check. The story given out was about a need to protect staff from a supposed Moscow viral infection. An internal state department memo said that 'considerable thought should be given to the establishment of a plausible cover for doing the tests because in the normal course of events . . . the press corps will learn of it, probably within two days of testing'.[10]

Extensive radiation studies on animals in *Project Bizarre* were undertaken by the defence department. A primate was exposed to microwaves at half the level permitted by the safety standard. A secret report concluded, 'There is no question that penetration of the central nervous system has been achieved, either directly or indirectly into that portion of the brain concerned with the changes in work functions'.[11]

The Moscow microwave affair came at an awkward time. During the course of the affair, defence officials in Washington told a congressional radiation inquiry that RF safety was already strict enough. The scientific evidence, they argued, did not justify the Soviet setting of a standard limiting exposure to a thousand times lower level. However, in the secrecy of his office, one official wrote that, 'for the record it should be noted that all the positive findings in the *Project Bizarre* were achieved at one half an order of magnitude below the accepted US standard for safe exposure'.[12]

Not until 1976 were the staff informed by the ambassador why they had been under medical observation for years. Their health, they were assured, was unimpaired. An unusually high white-cell count in some of the staff could be explained as a result of the Moscow virus. Nothing was said about the ability of microwaves to cause the same blood anomaly. Ironically, the microwaves were almost certainly beamed across the street to jam the embassy's own electronic eavesdropping. It is possible that the embassy staff had been working for years in a radiation environment contributed to by the electronic intelligence of both superpowers.

Though the Moscow embassy microwave affair was played out in an unusually exotic situation, it is indicative of official secrecy on radiation-related health issues. The reason is not hard to find: the military forces consider almost anything to do with radiation to be their business. Unfortunately, many scientists involved with radiation research and protection have gone along, either passively or actively.

Chapter 3

The microwave debate: how safe are the safety standards?

> *Dr Bill Guy* (expert witness): There is absolutely nothing in the scientific literature that would support claims of biological effects resulting from either short-term or long-term exposure to these extremely low levels of microwaves.
>
> *Commissioner Horsley's finding*: RCA may be willing to take [a risk] based on what they know and based on what they have to gain . . . The important thing here is not the scientific debate as to the extent and degree of present or long-term risk. What is important from a public perspective is the degree to which a broad segment of the affected populace perceives that they and their children are at risk.
>
> — *Decision on an appeal concerning a proposed satellite microwave installation at Bainbridge Island, Oregon, USA, 1981-2.*

IN Australia, it was not until the 1980s that microwave emissions became prominent as a public health issue. In some other countries many battles have been fought on the issue since the early 1970s. It was in Rockaway Township in New Jersey that a community won a classic victory against acclaimed radiation experts.

Rockaway Township is set in a quiet rural area about 50 km north-west of bustling New York. It is a retreat for those who wish to work in a metropolis and seek pastoral peace at the end of the work day. In 1980, Rockaway citizens found that a company planned to erect satellite dishes in the midst of their rural retreat. The dishes were to transmit microwaves carrying video programs to viewers scattered across the country. Satellite microwave relay of video, which is fed into local cable TV systems, is now big business in the United States.

The company went to great lengths to win the approval of Rockaway citizens by designing the installation so as to soften its visual impact. The zoning regulations covering the transmitter area had, of course, to be amended by the local authorities. That meant a public inquiry to deal with the rezoning application.

The company soon discovered that the real hurdle it faced was not the look of the dishes but parents' anxiety about microwaves being beamed over the local schoolground. Could radiation spill over from the dishes and irradiate the children? Certainly there would be some spill. But how much? What would be the effect on the children's health? These were the questions to which the community wanted definite answers.

To reassure the inquiry panel, and particularly parents, about the harmlessness of the operation, the company assembled some notable experts including Herman Schwan, biophysicist and US navy consultant. He was the author of the first US radiofrequency safety standard. After three nights of intense cross-questioning, it became apparent that Herman Schwan had failed to convince the inquiry that the dishes presented no danger to health. After months of other expert testimony the company beat a retreat from the town.

The company's expert witnesses failed in not being able to offer a water-tight guarantee that the microwaves would be harmless. The scientists could only say that they believed, on the basis of their research findings, that no one would be harmed. Of course, research was still going on. 'The company's guarantee,' said one local, 'doesn't mean a piddle hole in the snow.'[1]

The Rockaway community had been worried by what they were hearing about reported health effects of microwave transmitters in other places. The state of New Jersey is nicknamed 'cancer alley'; it has the highest cancer rate in the nation. The Rockaway community is one of the few still unaffected by the state's heavy industrial pollution.

Vernon is another rural community in New Jersey. Since 1974, when satellite dishes were first installed around the town, the incidence of birth abnormalities and leukaemia has risen well above national averages. Down's syndrome reached epidemic proportions, at five times the national average. In a pocket of one hundred houses there were forty instances of miscarriage, birth defect and nervous disorder. Not unnaturally, people looked to what had changed around Vernon and was likely to cause such an epidemic. They discovered that the rise of disease had begun about a year after the first satellite dishes were erected. Residents produced maps showing how the paths of beams from transmitters passed over areas of houses most affected by disease.

Experts explained quite reasonably to residents that an excessive number of sicknesses occurring in one area could be either a matter of chance or the result of some toxic agent. The agent, they said, perhaps less reasonably, could not possibly be associated with the dishes, because the microwave levels around the houses were well below the safe limits set by US standards. Vernon residents have remained unconvinced and have no doubt about the culprit: microwaves.

IN August 1985, even before AUSSAT sent Australia's first domestic satellite into orbit, protest voices could be heard against the siting of its earth stations. One station, intended to communicate business data, was planned for inner-suburban South Melbourne — an area already dotted with dishes. The new dishes would be poised over a housing estate, playgrounds, kindergarten, a medical centre, and a shopping centre.

Ironically, communities around the world are learning from each other about possible health effects of microwave exposure via microwave relays of public affairs programs. South Melbourne residents soon heard about the excess of disease in Vernon. Residents called on their councillors to refuse a permit to erect the dishes. Councillor Anne Fahey was indignant about the way the company and government seemed to think 'people are like white mice, you can do anything with them'. Certainly the choice of site displayed a singular lack of sensitivity to people's concerns about microwaves. After a heated council debate, the application to erect the dishes was rejected.[2]

Such conflicts with local decision-making do not always end as they did in Rockaway. Local communities in the United States seem to have more say in planning matters than do those in Australia. In the end, the company had its way in South Melbourne.

In Australia, the community experience of challenging urban planning is one of getting involved in tortuous hearings and appeal proceedings, against heavy odds all the way. Communities do not have access to the costly legal and technical advice available to corporations; they must rely on their own resources and abilities. They do not have the same easy access to government ministers as company and utility officials do when they plan projects with fulsome promises of economic growth. Even with the best will in the world, community activists are often exhausted by these struggles.

In their appeal against the council's decision the company erecting the South Melbourne transmitter held the trump card: the safe limits set in the Australian safety standard. Its installation would result in exposures well below the limit set by the standard. That, said local resident, Dr David Scholl, sounds reassuring, but 'the Australian standard is a pretty *ad hoc* thing and no one really seems to know what's dangerous . . . when you start talking about cancer and genetic dangers it's going to take twenty or thirty years to show up'.[3]

In Bunbury, Western Australia, another community battled against the erection of a TV dish. Information flooded into the city from all sides. The residents gathered information on the health effects of microwaves and put their case to the city council. The council approved the dish on the mayor's casting vote.

Here, too, the Australian standard was invoked to demonstrate that the irradiation of residents in the street where the dish was to be erected would be harmless. A government radiation protection officer explained to the city council that straying microwaves near households 'works out to be about one two-thousandth of the level allowed for public exposure'.[4]

What more needed to be said? Only that the standard was decided by controllers of radiation emitters virtually on the basis of what their transmitters could comply with. An unfortunate, silent aspect of the Bunbury case was that technically feasible alternatives were available. Residents had been denied a more socially acceptable solution. The standard, whatever its validity, had been used inappropriately to stifle the community's voice.

Marella Cottee lives opposite the new dish. 'We look straight into the face of the dish. It's quite sinister. It's quiet, it has no visible output, no sound and yet you know something is being emitted from the dish. That has a curious effect on people.'[5]

Councillor Judy Jones found herself unable to judge the merits of the conflicting opinions presented to the council. She had in the end to judge the experts themselves. 'If you look at the standard-setters and experts in the past then you can draw an analogy of women of my generation being affected by stilboesterone. Women died as a result. And yet the drug was approved by the standard-setters and the experts. What about the thalidomide epidemic? These might be examples outside the microwave issue but I make the point that government-imposed levels and expert opinion have been wrong in the past. Now, while we still have doubts, microwaves should be treated with a degree of caution.'[6]

Setting safety standards

Early in 1985 the Standards Association of Australia (SAA) launched its first standard for *occupational* and *environmental* exposures to radio waves and microwaves (in the RF band).[7] It is the standard quoted in the Bunbury affair. But, as councillor Jones said, standard-setters have had their facts wrong in the past. What considerations went into setting this particular standard?

The philosophy of the SAA is that the best people to set standards are those who have the relevant technical expertise and managerial experience handling the technology. The association accordingly recruited to its drafting committee mostly representatives of the military services, the electronic communications industry and allied professional bodies. Committee proceedings extended over four years. Only after drafting work was under way did the winds of change, bringing workers a say in their own safety, stir the SAA to open the door to trade union representatives. Community, environmental and health groups were still not invited in.

The maximum occupational RF exposure set by the SAA is

1 000 microwatts per square centimetre over an eight-hour-day for microwaves (1 000 microW/sq.cm); it is higher for radio waves (see box below). The Australian standard follows the 1982 US

SETTING PERMISSIBLE RF EXPOSURES

Two factors determining the energy that falls on the human body when exposed to RF waves are *frequency* and *intensity* of the radiation. The intensity of RF waves is measured as *power density*, which is generally expressed as *milliwatts per square centimetre* (milliW/sq.cm). Power density is also expressed as *microwatts per square centimetre* (microW/sq.cm).

Power density is a basic measure of the quantity of energy falling on the body. But not all the energy falling on a body is absorbed. The energy actually absorbed depends on frequency and is measured as a *specific absorption rate* (SAR) in watts/kilogram (W/kg). The SAR is not so readily measured and depends on the behaviour of the electric and magnetic fields generated inside the body, the size and position of the body, and other factors.

Because there can be hotspots the SAR will vary considerably over the body. However, the RF standard is essentially based on an averaging of exposure time and energy distribution inside the whole body and makes no allowance for the effects of hot spots. The SAR adopted for the Australian RF standard is eight W/kg over any one gram of tissue and 0.4 W/kg over the whole body.

Since heating by RF varies with frequency, *occupational* exposures are set according to frequency. For microwaves the level is set at 1 000 microW/sq.cm or ten watts/sq.metre in a working area. Bright metals focus microwaves, greatly reducing the margin of safety for workers. Less than half a watt focused on the human eye can cause cataracts. The exposure level for radio waves, which have a lower heating capacity than microwaves, may be up to 10 000 microW/sq.cm and still higher for short periods.

The permissible *environmental* exposure is one-fifth the *occupational exposure*: that is, 200 microW/sq.cm for microwaves but over a 24-hour period. This means people can be constantly exposed to a not insignificant two watts/sq. metre in an area where they live.

standard fairly closely. Until that year the level of exposure permitted in the United States had been ten times higher. But even the much-reduced level, set in 1982, is still forty times higher than in the Soviet Union and Eastern Europe (see figure 3.1 below).

What then is the scientific basis for the Australian safety standard? To find out, we must look to the history of RF standards in the United States.

In the 1950s, a committee headed by Herman Schwan, expert witness at the Rockaway inquiry, was requested by the US navy to determine a safe microwave exposure. From the start the committee dismissed the idea of adverse health effects occurring when there was no overheating of the body. That meant that they looked only for *thermal* effects.

A calculation was made of the exposure the body could withstand before incurring *irreversible* heat injury. This calculation

Figure 3.1: A comparison of the permissible occupational RF exposures in Australia, the United States and the Soviet Union

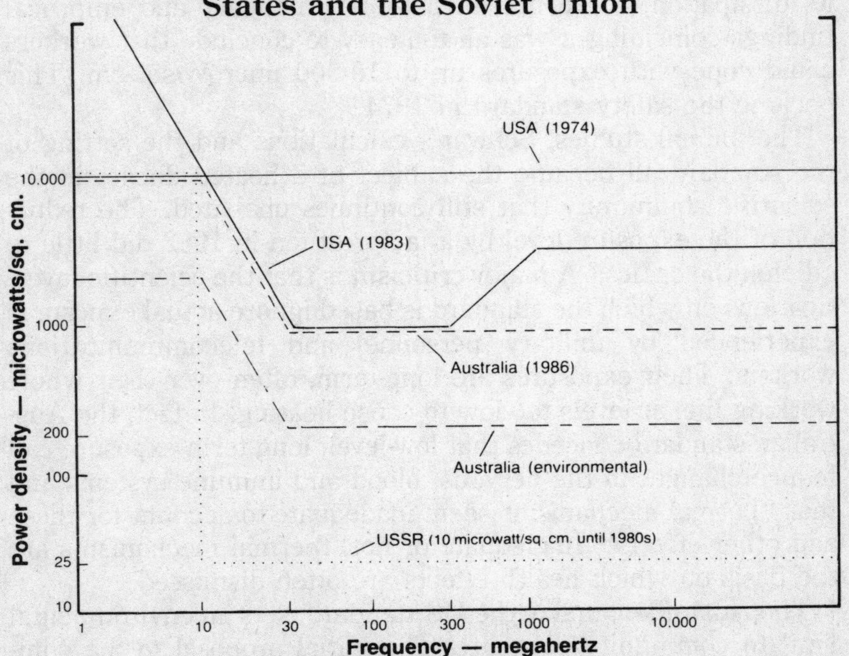

was based on two factors. One was the rate of conversion of microwave energy into heat in living tissue. The other was the ability of the human body to dissipate the excess heat. Both factors were studied, not by experiment, but on the basis of suppositions later shown to be flawed. Schwan personally made the calculation and arrived at 10 000 microW/sq.cm as the threshold below which microwaves would not cause irreversible damage.

Clinical studies on radar exposure of military personnel had shown evidence of eye opacities, headaches and blood abnormalities, but actual radiation levels causing these injuries were uncertain. Apart from these vaguely reported observations, hard data on microwave injury to humans was hard to come by.

Studies on animals were therefore relied on to assess health damage from body overheating. Dogs were cruelly 'cooked alive' to obtain data on microwave damage to eyes, testicles, glands and blood. It was concluded from these experiments that *permanent* injury occurred only above an exposure of 10 000 microW/sq.cm. This was the same threshold Schwan had calculated from consideration of the conversion of the radiation energy into heat and its dissipation in the human body. With theory and empirical findings coinciding it was all too easy to conclude that workers could cope with exposures up to 10 000 microW/sq. cm. This became the safety standard in 1974.

The animal studies, Schwan's calculations and the setting of the standard all became the subject of a heated debate in the scientific community that still continues unabated. The reduction of the exposure level by a factor of ten in 1982 did little to quieten the critics.[8] A major criticism is that the scientific investigations on which the standard is based ignore actual exposures experienced by military personnel and telecommunications workers. Their exposures are long-term, often over their whole working life, at levels too low to cause heating. In fact, the Australian standard concedes that low-level, long-term exposure can induce changes in the nervous, blood and immune systems and that 'thermal mechanisms seem inadequate to account for these and other effects'.[9] Inadequate or not, thermal mechanisms are the basis on which health effects are often dismissed.

The Australian, unlike the US standard, sets an environmental limit to community exposure. The initial proposal to the committee, put forward by CSIRO representative David Hollway, was 40 microW/sq.cm for microwave transmission. The proposal was unacceptable to industry and military representatives; 100 microW/sq.cm over a 24-hour period was adopted.

As the work on drafting the standard drew to a close, a letter from the Department of Communications was tabled pointing out that RF levels around Broadcast House in Adelaide exceeded the proposed 100 microW/sq.cm. Only then did the electronic media representatives admit that they could not meet the 100 microW/sq.cm limit. The standard, after all, was only a voluntary one. The safe level was increased to 200 microW/sq.cm over a 24-hour period to accommodate the actual exposures. Thus the integrated daily exposure of a resident living near a microwave transmission tower is allowed to be nearly as much as the integrated occupational exposure over an eight-hour day. Yet, according to David Hollway, 'the lower value could be adopted without interfering with the present use of the RF spectrum'.[10]

The secretary of the SAA said, 'We know of no reason why anybody should be worried . . . the evidence we have is that less than 0.5 per cent of the people in Australia will be affected'.[11] Even this small percentage means 75 000 people can be exposed. Transmission dishes and antennas are still proliferating. In the absence of more prudent siting, the number of people living and working close to RF sources will increase. But the numbers are not the main point; the need for protection should not be dismissed just because a community is small. And, when communities try to keep sources of radio waves and microwaves at a distance, they are told the emissions will be only a fraction of what is set in a scientifically based safety standard.

IF it had been intended that the setting of the Australian standard would avert the kind of heated public controversy that occurred in the United States, it did not turn out that way. Almost before the ink had dried on the document, it became the centre of controversy. The Australian Council of Trade Unions (ACTU) withdrew from the committee and refused to endorse the standard, on the grounds that it did not accord with the most recent research findings on athermal effects.

One view of standard-setting, according to David Hollway, is that 'we should "play safe" by setting low levels now and then raise them only if later research shows higher levels to be harmless. This view usually appeals to those who are actually being irradiated in the course of their daily work'. Or, it could be added, while living in their residences.

'The opposed view,' David Hollway says, 'is that the level of radiation which everyone agrees causes demonstrable harm, should be found as accurately as possible and the permitted level set at a not-too-large factor of safety below the danger level. This view has more appeal to those owning or controlling sources of radiation. Eloquent claims are made that this is the only "scientific method" of setting maximum exposure levels because they are based on proven facts. My view is that, far from being scientific, this procedure is unintelligent at best and often is disingenuous'.[12]

Disingenuous perhaps, but stricter health and safety standards mean extra costs for electronic and allied industries. Military services do not want standards to hamper defence capability. 'It is reasonable to err on the safe side,' a consultant said to the electronics industry, 'but not so far that it hurts; not so far that progress in the art becomes jeopardised'.[13]

ALL technology carries some risk. According to Tony Morgan, an industrial hygienist with Telecom, 'Whenever you set up a standard, you have to take a risk based on existing knowledge'.[14] It is the price society must pay, we are often told, for technological benefits. The other side of the coin is workers' safety. It is their health, not that of the standard-setters, that will be impaired if a 'safe' level of today turns out to be unsafe tomorrow.

Sandy Doull has had a long experience with RF safety issues as a trade union health and safety research officer. He says, 'It costs money to protect people properly; quite a lot of money. Protective clothing is available but it is not cheap, and to re-engineer equipment is not cheap either. For as long as they can officially deny there is a problem they can get away from paying for large costs to prevent injuries and get away also from paying compensation to the workers who are affected, but don't know that their disease comes from this source.'[15]

Implicit in occupational safety standards is that workers may be exposed to more radiation than other people; that to earn their money workers must accept higher personal risks. Apportioning risk in this way is fundamentally flawed. Vocational choices are not made so freely as is commonly supposed. People's job opportunities are limited by their socio-economic position and educational opportunities. At one time workers were resigned to the conventional inequities of risk-taking; they bargained for danger money. This attitude to work is changing. Today, unions insist on the elimination of work hazards by being part of health and safety supervision.

Community health and environmental groups do not yet have an effective voice on exposure to radiation. Exposure remains involuntary. The community can intervene in the process of considering a new, perhaps unwanted radiation-source only through submissions to government review-processes — and then only in the final, not conceptual, stage of a project.

The accumulation of data needed to estimate a radiation risk has to be based on a willingness to support the relevant research. That support has not always been forthcoming. The research can be costly; but the public health cost of a technology's emanations, hidden though they usually are, can be enormously greater.

The data generated by environmental and health studies depend a great deal on the questions asked. The choice of questions is influenced by the preconceptions of those with a finger in the research pie. Research can be well designed, carried out competently, the findings clearly presented — and yet do nothing to advance knowledge in the field. Dr Rochelle Medici, a researcher on animal behaviour, says, 'It is as though scientists had retreated from doing challenging, frontier studies because such work engendered too much controversy or elicited too much criticism. We are left with "safe" but meaningless experiments. The results of such experiments are a foregone conclusion'.[16]

If people are to have wider participation in setting safety standards, then environmentalists, community health workers and trade union safety officers must have the opportunity to put pertinent questions to researchers arising from their experiences with radiation sources. However, while community concerns need to be addressed, the design and execution of research projects will necessarily be in the hands of scientists. Research projects of the kind that investigate the chronic effects of weak EMR can take years and may be very expensive.

'A layman reading a scientific report sees an extremely orderly and precise final report,' Rochelle Medici says. 'However, that report is the result of a disorderly and sometimes highly intuitive series of pre-experimental decisions and plans . . . These decisions depend on such things as the scientist's awareness and interpretation of published literature, his or her own training, one's responsiveness to the needs of the granting agencies, current fads and fashions in science, the personal foibles of the scientist.'[17] There has to be trust on both sides. Unfortunately, public trust in those responsible for setting radiation safety standards has been slowly eroding.

Thermalists v athermalists

There are two camps in the microwaves debate — the *thermalist* and the *athermalist*. During the late 1970s there was rising criticism by athermalists that research projects financed by the military services were ignoring low-level effects. Researchers were using such very high radiation intensities that any athermal biological changes would be confused with thermal effects.

The US air force finally agreed to finance a large research program designed to address questions about athermal effects. Dr A. (Bill) Guy, an engineer, not a biologist by training, was appointed to conduct the research. The primary question addressed was, 'What is the effect of protracted exposure to weak microwave radiation?'. The exposure decided on was 21 hours a day over 25 months with microwaves at 480 microW/sq.cm, giving a specific absorption rate of 0.4 watts/kg on which the US RF safety standard is based (see box: Setting permissible RF exposures). One hundred rats were exposed and another hundred were used as controls.

A second question addressed was one that athermalists had been asking for many years. It was: 'What if the radiation, instead of being in the form of a plain *continuous* wave, has a *modulated* or *pulsed* waveform?'. Athermalists believed that pulsed and modulated waveforms were more biologically active.

We can picture modulated waves as having ripples superimposed on them. Radiation is pulsed, simply by being turned on and off, like flickering light from an electric bulb. The microwaves used in Bill Guy's experiment were modulated at 8Hz and pulsed 800 times a second.

Pulsed and modulated microwaves were tested for two reasons. One is that artificial modulated and pulsed microwaves now permeate our environment. Radio waves are modulated to bring us music and speech on our radios; radar uses pulsed signals. The other is that the modulation and pulsing frequencies of artificial EMR are in the same band as the natural ELF fields in our bodies. If living cells have evolved 'tuned' into this frequency band, would they have, it was asked, the facility to demodulate the artificial waves? If so, then it would be the modulation or pulsing frequency that was biologically significant and not the carrier wave frequency that counts.

Guy's research took five years to complete. At the 1984 annual meeting of the Bioelectromagnetics Society he reported finding severe health damage in the rats.[18] A high incidence of malignancies and benign tumours in the endocrine gland system was observed.

The observed growth of benign adrenal tumours in the exposed rats could help to explain frequent complaints by radiation workers of headaches, dizziness, memory loss and fatigue. These symptoms have become known as 'microwave sickness'. Benign adrenal tumours, or pheos, are associated with high blood pressure and headaches. It is an example of how 'subjective' complaints can be dismissed out of hand, only to find support later in findings of scientific investigation.

The reception of the results says something about the play of influences on deciding research topics. Malignancies were not what the researchers had expected; they leaned towards the thermal hypothesis. Bill Guy had previously testified at public inquiries that there was no evidence for athermal effects. A scientist who reviewed the research for the US air force commented that 'the finding of the excess malignancies in the exposed animals is provocative'.[19] Unquestionably so. If confirmed, the findings undermine existing Western standards for RF exposure, including the Australian standard launched only in 1985.

In fact, malignancy resulting from pulsed microwaves had been reported as long as 20 years before. In 1962, Drs Prausnitz and Susskind observed increased neoplasm of white blood cells.[20] However, this earlier work was complicated by the use of high-level microwaves. It shows how, in research, looking for certain effects can influence the design of an experiment in a way that clouds other effects. It illustrates the benefit of outside questioning of specialist research directed towards establishing the scientific basis of radiation health and safety standards.

Guy's findings, of course, need further investigation before firm conclusions can be drawn on the effect of modulation. Unfortunately, it seems history is about to repeat itself. The US air force is publicly downplaying the results of the Guy experiment and does not intend to sponsor follow-up research. It is hard not to think that air force officials were expecting an 'all clear' for RF emissions.

Chapter 4

From submarines to power lines: The extra-low-frequency debate

Question: Do you believe power lines out in our community in Australia are posing no potential threat to health at all?

Answer: Well, I believe that if they conform to present recommendations in the WHO document on this subject [exposure to power-line radiation] that they probably won't be causing any harm to people on the present basis of our knowledge.

Question: So far as you know?

Answer: You always have more research to be done . . .

— *Interview with Dr Repacholi, chairperson, World Health Organisation (WHO) review of electric power and community risk,* Four Corners, *Australian Broadcasting Corporation, 2 September 1985.*

HIGH-VOLTAGE cables draped between tall pylons are a familiar sight on the fringes of our cities. These cables come from power stations in country regions. Once in the city the cables enter distribution substations, where their voltages are transformed to lower levels. Distribution lines radiate out from substations along gullies and roadways. The open spaces along the urban easements beg to be turned into playing fields.

Since the 1880s, when the first power stations were built, power-line voltage has climbed steeply. By the 1920s it had risen to 100 000 volts. Today in Australia it has reached 500 000 volts and, overseas, 750 000 volts. High-voltage power lines now criss-cross most of the earth's land surfaces.

It is generally pictured that electricity flows inside the cables like water through a pipe. However, the electrical energy actually flows in and around the cables. Invisible electric and magnetic fields adjacent to the cables oscillate to and fro along the length of power lines at the same frequency as the alternating current. In Australia the frequency is 50Hz (see box below).

With *extra-low-frequency* EMR, or ELF radiation, each field is measured separately. The field strengths fall off sharply with distance from the lines. Nonetheless, fields 40 metres away from a 500 000 volt line can cause an ordinary fluorescent tube to glow. A person standing under a power line is immersed in electric and magnetic fields. Around the curved contour of his head the electric field is about 15 times stronger than in the space around. Under power lines we wear an invisible electromagnetic halo. High on the pylons, workers can have their hair stand on end; on foggy nights workers' heads can assume glowing halos.

Electric and magnetic fields interact with the human body.

ELECTRIC AND MAGNETIC FIELDS

Whenever an electric current flows along a wire an *electromagnetic field* emanates into the surrounding space. An electromagnetic field is composed of an *electric* and a *magnetic* field travelling at right angles to each other.

These fields are only imperfectly understood. They are a convenient way to describe the 'action at a distance' of electrical and magnetic forces. We can visualise a field and its lines of force from the pattern of iron filings surrounding a magnet.

Direct currents emit *static* electromagnetic fields. *Alternating* currents emit fields oscillating at the frequency of the current. These fields have *extra-low frequencies* (ELF). In Australia and Europe the frequency is 50 cycles a second (50Hz). In the United States it is 60Hz. A recent trend overseas, but not yet in Australia, has been towards using direct-current transmission at very high voltages. This will change the radiation environment around the high-voltage power lines from oscillating to static fields.

At distances ELF waves behave as radiating waves travelling through space like other RF waves. The military use special antennas to broadcast ELF waves to submerged nuclear submarines.

Near high-voltage sources the body is weakly coupled to the electric field. A small electric current, generated inside the body, flows through the feet to the ground. The magnetic field permeates the human body more uniformly than the electric field, and sets up eddy currents in body organs.

Because children's bodies are smaller, the currents leaking through their bodies are weaker than for adults. However, children's bodies have many growing cells and they are developing rapidly, both physically and mentally. This makes them more likely than adults to be affected by the fields. At some schools children play and exercise within reach of the fields. The induced body current increases when the arms are extended during play, causing the field strength in the body to fluctuate.

A vehicle parked under power lines collects a large electrical charge because it is insulated by its rubber tyres. Farmers

The *strength* of an electric field associated with a power line is measured as *volts per metre* (v/m) and is proportional to the line's operating electrical potential (voltage). The highest field-strength likely to be encountered on the ground under a power line will be about 10 000 v/m. This is millions of times greater than the *natural* electric field at 50Hz.

When an electric field is coupled to an object it induces a current in it. Metal objects can become highly electrified. The human body is only weakly coupled; the small currents flow to the ground. Body currents are measured as *milliamps* or *microamps*.

The strength of a magnetic field associated with a power line is usually measured in *milligauss*. The field is proportional to the strength of the current flowing along a power line and its *configuration*, which involves such factors as the distance between the active and return lines and how near to equal are the currents in the lines. In household cables the active and return wires are close and balanced in current flow, and so have extremely low field-strengths.

The power-line magnetic field will generally be weaker than the natural geomagnetic field. However, the natural field is *static* while the artificial field *oscillates*. This has considerable biological significance.

ploughing around power lines must make special provision for their large tractors to discharge the electricity they pick up. Otherwise, anyone touching the vehicle may get a sharp and even painful shock.

People living in houses close to or alongside power lines have seen their fluorescent light tubes glow in the night even when the switches are off. In these extreme situations houses have to be shielded from the electric fields. But weaker fields pervade houses and the people living in them, unseen and unfelt.

For a long time the likelihood of health damage to people living exposed to power-line electric and magnetic fields was generally dismissed. The energy levels of the ELF waves are too low to heat body tissue. According to the thermal hypothesis, there can be no serious health damage if there is no heating.

The possibility of adverse health effects from exposure to ELF radiation was first raised seriously in the early 1970s in a report by a committee asked to examine the environmental impact of an ELF communication project proposed by the US navy. The report was not made public.

Sending signals to submarines

Because nuclear submarines are submerged for long periods they present a special problem in radio communication with their home base. In the 1950s it was discovered that, because ELF waves lost little energy as they travelled through water, they were ideal for communicating with submarines wherever they happened to lurk in ocean depths.

The first small test-ELF antenna operated successfully in 1971. The navy then planned to build a full-scale antenna, code-named *Sanguine*. With the technical problem solved, the navy encountered another problem: community resistance to living right inside an ELF antenna. The *Sanguine* antenna was to be constructed from 2 000 km of electrical cable laid out like chicken wire and buried under 10 000 sq. km of land. People would be shielded from the *electric* field by the ground. But the *magnetic* field would permeate people living within the antenna.

Sanguine was to transmit at a frequency of 76Hz and a wavelength of 3 900 km. The electric field oscillates almost vertically to the earth's surface, with the magnetic field parallel to it. These extraordinary 'waves' use the space between the ionosphere in the upper atmosphere and the earth's surface as a waveguide. At the ocean surface the ELF waves go down to the submarine and are received by a long-trailing aerial.

Sanguine was to be built in the state of Wisconsin in an area where the earth had the required low conductivity. The now familiar pattern of conflict developed between community and authority. Environmentalists campaigned to stop the project; the navy sponsored a review of ELF health effects, as part of a 'compatibility assurance program'. The very name begged a safety clearance.

The review committee disappointed the navy by reporting that biological effects of ELF radiation had been demonstrated in a number of laboratories around the world. The occurrence of bioeffects is indicative of body stress but does not necessarily signal lasting health damage. The human body is capable of recovering from many kinds of stresses. However, the report suggested that lasting health damage must be considered seriously wherever people were exposed to ELF radiation for a long time. That raised the wider question of the health of large numbers of people living alongside high-voltage power lines.

'The committee unanimously felt', committee member Robert Becker reflected later, 'that major segments of the civilian population were currently at risk'.[1] The wider risk to public health from power lines created a dilemma for the committee, appointed to consider only the naval ELF radio station. They recommended further investigation into the health effects of ELF radiation.

The navy promptly tucked the report out of sight. A navy statement proclaimed ELF to be 'no problem' for public health. Inconvenient information on radiation can be as invisible to the public eye as radiation itself. But in 1975 a Wisconsin senator unearthed the report and promptly made it public. The navy retreated to more remote country to build a trial antenna.

Years of unabated opposition in Wisconsin led eventually to abandonment of the *Sanguine* project. Instead the navy announced its intention to build a smaller above-ground antenna renamed *Project ELF*. In 1984 an appeals court ruled that, health damage or not, *Project ELF* must proceed. Otherwise, the judge said, the United States would trail the Soviet Union in ELF submarine communication systems. The cost to national security would be too great (this is another instance of how national security can perhaps be insisted upon at the expense of the 'protected' more than the 'enemy').

After a decade of controversy and the green light from the courts, the navy still wanted to keep on the right side in its public relations. It again enlisted the experts, this time asking the American Institute of Biological Sciences (AIBS) to set up another committee to review ELF health effects. The navy seemed confident that a report coming from such a reputable institute would put the health issue to rest.

The AIBS committee did as the navy hoped and reported that 'exposure to ELF electric and magnetic fields in the range of those produced by the ELF communications system does not pose public health problems'.[2] There was a time when such an authoritative finding could be used to quash any attempt to raise the issue again publicly. Not so in 1985. Reports on controversial environmental issues from government-sponsored experts are now treated with some scepticism.

Scientists, like most of us, have theories about the way the world works, and are inclined to prefer the evidence that fits their theory. So when anomalous results turn up that do not fit the theory, the natural tendency is to explain away the results rather than question the theory. Again, like most of us, scientists can have strong feelings about what is best for society. However, today, scientists generally work in highly specialised compartments. Being expert in this or that speciality does not, in itself, equip a scientist to deal wisely with broad environmental questions. For these we need not only scientific expertise but a broader knowledge.

The Wisconsin Department of Justice appointed a committee comprising representatives with legal, medical and radiological backgrounds to look at the AIBS report. The committee was a commendable mix, lacking only a socio-environmental perspective.

The committee prepared a *Commentary* which noted a number of inconsistencies.[3] Radiation hazards described in the body of the report, 'had been edited out of the report summary or had been rewritten so as to substantially change their meaning'.

Studies that came up with findings showing no bioeffects from ELF fields were treated more gently than studies positively demonstrating bioeffects. Findings of adverse health effects were 'rigorously critiqued throughout the report, whereas experiments with null results received less thorough review. It is customary in evaluating health effects to place more emphasis on positive results'. Empirical evidence had been dismissed simply 'because current theory is unable to match a mechanism of action with the observed phenomena'. And the objectivity of the AIBS committee was questioned: 'Many members of the committee receive substantial funding from utilities or the Department of Defense', causing a potential conflict of interests.

ELF radiation in the schoolground

One night in March 1986, at a lively meeting in an inner suburb of Melbourne, a speaker waved aloft a heavy-looking official report like an evangelist might display the Bible. It was the AIBS report. The speaker was its compiler, Professor of Biology and Poultry Science at the State University of Pennsylvania, H.B. Graves.

Graves had been brought to Melbourne by the Victorian Health Commission to give expert advice on the health effects of two power lines. One was proposed for the Merri Creek, passing through inner suburbs. The other was an existing power line in an outer suburb. Graves explained how his job was to interpret the scientific data. He would apply the same high standards of scientific objectivity, he stressed, as he had when compiling the AIBS report.

In 1980 the State Electricity Commission of Victoria (SECV) proposed a 220kV line be built along the Merri Creek gully. For many years this gully has survived the inclination of engineers to turn the city's gullies into concrete gutters or freeways. Today, the gully's open spaces serve as wooded parkland and recreation fields.

From the start the proposal ran headlong into opposition from residents determined to save the gully's natural setting. However, by the time the SECV published the environmental effects statement (EES), in 1985, residents had found other reasons to oppose the power line. They had learned of the possibility of ill effects of electromagnetic fields. The power line either skirted or passed over three schools.[4]

A 500 000-kilovolt power line runs alongside Rangebank primary school in the outer-Melbourne suburb of Cranbourne. Parents, once they learned about the electromagnetic field in the playground, sought to have the power line redirected away from the school. Some parents placed their children in another school. The school exercise-area was moved from under the power line.

The power line was there first. But by the time the school was to be built, health questions already hung over the SECV's power lines. The SECV could have informed parents about the health issue, confident though they may have been that the children would not suffer. An alternative site could have been chosen. All the fuss and bother would have been avoided. Parents have not only failed to have the line removed, but the SECV has now flanked the school with another line.

Rangebank parents were not alone in worrying about power lines skirting their children's school. 'Almost every report we read,' said Don Collins, principal of Klein Oak school, in Houston, Texas, 'and it didn't make any difference the authorship — there was an indication that additional research was needed. And they said more time is needed to determine the potential effects [of ELF fields] on students. At that point we firmly established with our board of trustees that we were not going to allow our students to be used as subjects of research'.[5]

In 1981 the Klein Oak school board unsuccessfully opposed rezoning to allow the power line to be built. In 1984, with the project completed, the power was switched on. In 1986 a court ordered that the current be switched off. A jury found the utility had shown a 'reckless disregard' by erecting a power line where it could impair the students' health. The utility was ordered to shift the line and pay $25 million to the school in compensation. The company shifted its power line. A higher court quashed the compensation order.[6]

The utility

Australian power utilities are cast in the technocratic mould. They are highly dedicated to building their electricity supply systems. They feel, quite rightly, that their engineering achievements have gained wide public approval. Popular perception for a long time has been that energy growth is a good thing for the economy. Energy growth, especially electrification, is associated in the public mind with progress.

Nevertheless, in recent years countries have demonstrated it is possible to produce more with less energy. Moreover, as we learn about the environmental and health effects of energy use, we are given another good reason to want to conserve energy. All this is leading to a re-evaluation of the desirability of energy growth. But the utilities are finding it hard to adapt to changing public perceptions.

When the Collingwood Residents Association objected to the Merri Creek power line, the SECV's response to the community's environmental and health concerns was blunt. Whether or not to build the line was a choice between a trumped-up health risk and a genuine risk of a brown-out in the city. The consequences of not building the line would be intolerable. 'Thousands of people could be stranded on public transport and in lifts.'[7] The cost of putting the line underground would burden all electricity users with higher tariffs and 'electricity customers are fed up with paying higher power bills'.[8]

However, the SECV proposed that, in view of public concern about the health effects (whipped up by the media), an independent expert should be invited to Australia. The expert's brief was not so much to advise the SECV on the health effects of its lines as to 'clarify the situation for the community'.[9] The expert the SECV had in mind was Graves.

When the health dispute first blew up, a public meeting called by the residents association proposed an open inquiry to sift the scientific evidence, in which all protagonists in the dispute, including of course the local communities and councils, should participate.[10] After all, workers participate in setting safety conditions in workplaces; why not community participation when off-site pollution affects residents?

The state minister for health went so far as to agree that 'it is clear that the scientific basis for the community concerns needs to be examined'. But how? In the minister's view, 'it is desirable that the opinions of an independent scientist skilled in this area be obtained'. Also it was important that the 'public consultation processes envisaged under the Act have been fully explored'.[11] That meant no more community involvement other than written submissions on the EES. Not made public, and only later revealed under Freedom of Information, was a decision that SECV officers 'liaise with the Victorian Health Commission to arrange for a suitable expert'. The SECV consulted with the National Health and Medical Research Council (NHMRC) to confirm that 'Dr Graves is a suitable expert to undertake the proposed review'.[12] The SECV was to pay $120 000 to 'clarify public thinking'.

'I think we already know what he's going to say,' resident Andy Green said. 'We know where he stands in the scientific spectrum and it's a foregone conclusion as to what his conclusion will be. To be really democratic you would have a spectrum of scientists available to public scrutiny and feedback.'[13]

Out of self-defence, communities which have unwanted technologies thrust into their region search out their own information about potentially harmful emanations or effluents. Science news magazines and electronic science programs have become key information sources on topical environmental hazards. These sources often reveal diverse views among scientists when official handouts are uniformly reassuring.

Environmental groups now disseminate information along national and international alternative media networks. Talking across professional-lay barriers is building bridges between community researchers and those scientists more in tune with changing environmental values.[14] At no time did the Victorian radiation advisory committee or radiation protection agencies ever provide anything remotely like a balanced review of all the relevant findings on health effects. It was left to SECV engineers to provide the public with a fairly selective review of the biological effects of ELF radiation.

Faced with an expert who they believed had a conflict of interest because of his association with the utilities and the military, the residents sought a second opinion from a scientist whose views represented another part of the scientific spectrum. They invited Dr Jerry Phillips, Professor of Biomedical Research at the Cancer Therapy and Research Foundation in Texas, to Melbourne. His brief was to provide the public with information neglected by the government agencies. Jerry Phillips carries out research into the effects of ELF radiation on human cancer cells.

The minister for health assured residents that any advice Phillips had to proffer would be taken into account when the government assessed the EES. The government stretched its original guidelines to accept public comment on the Graves report. Was this to be democracy in action? The government seemed prepared to listen to the two sides, though on one side was the SECV Goliath and on the other was the community David — with an obviously unequal share of finance, influence and power.

Predictably, Graves's report gave SECV power lines a clean bill of health. Or almost. It came to the 'firm conclusion' that the strength of the electromagnetic fields of the SECV's power lines were 'very unlikely' to lead to adverse health effects.[15] The government seemed not to notice Dr Graves's muted ambivalence, announced it was satisfied on health and safety, and approved the project. But had the Victorian health department made up its mind long before?

'An awful lot of papers have been written which talk about these harmful effects but none of them have [*sic*] been scientifically validated', an agitated public health official burst out angrily at a public meeting. 'The only thing that would worry me,' the official went on, 'is not the hazard but what our [health] department does about [TV] programs like *Four Corners* and its exhibition of bogus science and the effect it is having on our community.'[16]

It was the second public meeting called to allow Dr Graves to explain how he intended to go about investigating the fields from the two SECV power lines. He had supposedly yet to reach any conclusions about the safety of the lines. Evidently, the health department's only problem was how to deal with public misconceptions about ELF radiation.

People who had made submissions on the EES knew by now that their submissions may as well have gone straight into the office paper-shredder. So too, they thought, could the expensive report which the expert had yet to write, for all it had to do with deciding the issue. The only avenue remaining was direct action.

BY early 1988 the political tide had turned strongly against the line. The Victorian Trades Hall Council and the eight municipal councils along the route all opposed the siting of the line for health and conservation reasons. The climax came with arrests of protesters committing acts of civil disobedience to stop the erection of pylons in the gully.

The government called a halt to construction of the line, and set up a powerline review panel headed by the Commissioner of Environment, David Scott. In a more enlightened way than had ever happened before, the panel set out to identify 'the most effective means by which people can be assisted in the consultative processes of the panel'.[17] The panel began by creating a pool of information on the issues and options, to serve as a basis for informed discussion with all protagonists. It was the kind of open consultation that the Collingwood Residents Association had proposed in the beginning. The panel's terms of reference did not include the health issues, but nevertheless these issues were raised from all sides.

Remarkably, the panel was able to produce a summary of the opposing positions agreed to by all protagonists. The agreed summary stated, 'The epidemiological studies have tended to generate either of two responses from scientists, engineers and lay people. There are those who argue that associations and causal relationships have not been conclusively demonstrated. There are others who interpret the same results as providing legitimate grounds for concern and caution, especially with respect to public policy response'. A point of agreement was that there 'is the need to continue with further well-designed epidemiological studies'.[18] The panel concluded that the most environmentally acceptable solution was an underground cable along a commercial road. The proposed route was away from residential areas, undoubtedly as a response to public concern about health and the need for caution. The panel's prescriptions for environmentally acceptable power lines was for new projects only. It recognised the mammoth cost of moving underground the hundreds of kilometres of existing lines criss-crossing urban areas.

What then of other residents, and students at Rangebank and other schools, still living and playing in the ELF fields of existing power lines? People's perceptions of intangible qualities were a crucial factor in the preservation of Merri Creek. However, perceptions of our environment are formed not only by seeing objects, but also by being conscious of their qualities. Science is giving us new perceptions of environmental quality. Knowing that an electromagnetic cloud permeates our children's school can cause as much distress as seeing a power line draped over a gully. 'Knowing you are permeated by EMR,' Marella Cottee said, 'has a curious effect on people'.

'When I see a line existing that close to the [Cranbourne] school,' Jerry Phillips said, 'I am quite displeased. I don't consider that a prudent solution to the movement of electricity from one place to another.'[19]

Whatever the risk of ELF radiation, children will be the most vulnerable. Children who have been exposed at school and home will be apt subjects of future epidemiological research. Presently, the Victorian government is committed to a program of placing lines underground to protect valued streetscapes from unsightly overhead cables; it is not acting to protect the health of children, even in the face of uncertainty about the harmful effects of ELF radiation. Perhaps Klein Oak school principal Don Collins had the right idea when he said, 'We were not going to allow our students to be used as subjects of research'.

Power-line radiation research

Epidemiologists search for connections. They use statistical methods to evaluate possible associations between occurrences of diseases in populations, and environmental, dietary, or other potentially harmful causal agents. An epidemiologist may identify a likely disease group in a locality or workplace by searching a cancer registry or by analysing death certificates.

However, before any evaluation can be made of a potentially harmful agent, the disease group must be matched carefully with a control group — possibly from census records — for factors such as socio-economic standing, age and gender. By careful matching of the two groups, epidemiologists try to cancel out variables that would bias the findings. For instance, better-off people are likely to survive the onset of a disease longer than people living in poverty: this tends to reduce the number of deaths from the disease among the better-off over a certain time span. People also vary, according to age and sex, in their susceptibility to agents which cause cancer

The design of epidemiological studies presents problems, as there are generally recognised limitations to the design of any study. Personal perceptions, as well as numbers, play a part. Having a hunch may turn up useful evidence of the causal agent, but it may also mislead. Obviously, there is much scope for scientific disputation. When the findings become a challenge to ingrained social habits or entrenched vested interests, the uncertainties inherent in the epidemiology will inevitably be used as fuel in a protracted and heated debate.

Epidemiological studies can only point to the probable causal agent. However, 'to be probable' means 'to be reasonably likely to happen'; and when that involves a serious disease, people have a legitimate case to argue for protective measures by health and environment protection agencies. The conviction that smoking caused cancer grew over decades and as a result of many scientific studies. Now a lesser relationship between passive smokers and cancer is beginning to show up. A long time passed before governments took measures to restrain smoking in public places. Despite all the epidemiological evidence to the contrary, the tobacco corporations have stuck staunchly to their claim that there is no demonstrable mechanism linking cigarette smoke and cancer.

Something of the same kind of defensiveness is displayed by the utilities, which have refused to entertain any possibility that their product is a potential health risk. They simply point to studies showing negative findings. Of course, studies on ELF radiation at present offer much less positive evidence of harm to health than do studies on cigarette smoking. And no one suggests banning electricity. But it is reasonable, on the basis of existing evidence, for a community to suggest that utilities should take all available measures to minimise ELF radiation exposure of residential areas.

'This is a genuine scientific debate on human health,' said Hec Gallagher, whose home would have been under the shadow of the now-abandoned Merri Creek power line. 'I can't say for sure there is a health risk so I can't ask them to take down the line in front of my home. But they can't say for sure there is no risk, so they have no right to put up another one over my back fence.'[20]

IN the late 1970s, Drs Wertheimer and Leeper undertook the first residential study on ELF radiation. They surveyed people living near power lines in Boulder and its neighbour city, Denver. They found more childhood and adult cancers in homes irradiated with ELF electric and magnetic fields emanating from power lines than in an unexposed control group. Families at greatest risk were those living within 40 metres of power lines and where the current flow along the lines was highest.[21] Wertheimer and Leeper concluded that the excess cancer occurred where residents were exposed to fields of about 1 milligauss or higher. Just how weak are such magnetic fields can be seen by comparison with the earth's geomagnetic field: it ranges from 250 to 700 milligauss. However, the geomagnetic field is *static* whereas power-line magnetic fields *alternate* at 60Hz (50Hz in Australia); this alternation of fields appears to have a potential for inducing biochemical activity in body tissue cells.

It is a difficult task, when studying long-term ELF residential exposures, to estimate the strength of magnetic field to which people living in a house, over say the past ten or twenty years, have been exposed. The Wertheimer research team broke new ground by using a coding scheme for *retrospective* estimation of very weak residential magnetic fields. The code was based on the 'ground' currents that flow to earth from house wiring, commonly through a water pipe, and the *configuration* of power lines skirting a house. A magnetic field is created by an imbalance of fields produced by currents flowing in opposite directions along the power-line cables; its strength is dependent on the spacing arrangement (configuration) of the cables. Since power-line configuration generally remains unchanged over the years, and ground currents can be measured, they offer the possibility of estimating the level of magnetic-field exposures over long occupancies.

Other research teams in the United States looked at the wire-coding method and found it to be a plausible basis for estimating exposure levels existing in past years; also, they concurred with Wertheimer that household electrical wiring and equipment did not contribute significantly to these residential fields.[22] However, wire coding gives only approximate estimates of field strengths as they were in the past, and estimating dosage remains a difficult problem for epidemiologists studying ELF radiation. As well, wire coding is applicable only for certain wiring systems, such as when the return current is grounded; if not, the magnetic field will depend mostly on power-line configuration.[23] Curiously, it is only since Wertheimer's studies that the *minimisation* of residential magnetic fields has been seriously considered. If households were being pervaded by a chemical that could be sensed, efforts certainly would be made to keep the pollution to a minimum. The SECV has done little more than contend, against the weight of evidence, that normal domestic electrical

ELECTROMAGNETIC FIELDS SURROUNDING ELECTRICAL EQUIPMENT

The fields surrounding working electrical equipment in the house can be extremely high; however, they hug the casing of the equipment pretty closely, and generally no one stays close for any length of time. When we are in close proximity our hands and arms are enveloped in strong fields. Unlike power lines, most equipment behaves as a point source; when we move beyond arms' reach the fields are negligible.

There are two exceptions where exposure to the fields may be prolonged:

— in the electric and magnetic fields around a 'live' electric blanket;

— in the magnetic field around a hair blower.

Because of the way electric blankets and hair dryers are often used, sensitive body organs are exposed to the relatively high fields for significant periods of time.

Exposure also may be prolonged when one is sitting at the screen of a computer monitor (the TV screen is normally viewed from further away). The magnetic fields are generally not so high as some other equipment; but some people — including the young when they are word processing for their student projects or playing games — often sit in front of a display screen for hours on end. The highest fields are at the sides and rear of a monitor.

usage contributes as much to the *ambient* fields in a house as closeby power lines[24] (see box below).

A Swedish residential study found an excess of cancer among people living near power lines and exposed to magnetic fields of more than three milligauss.[25] Studies on the health of workers in electrical workplaces have indicated excess cancer, over either the national norm or other occupational groups. These occupational studies generally suffer from a lack of proper matching of exposed and unexposed workers; nonetheless, taken together, the findings are suggestive of an association between ELF fields and cancer.[26] Other residential and workplace studies have failed to find a connection between ELF radiation and cancer.[27]

The most extensive investigation yet undertaken on the biolog-

Wires carrying current to and from a house are close together; the current is balanced and fields emanating into rooms are negligible.

Electrical Equipment	**Magnetic Field** milligauss **Against case**	**Magnetic Field** milligauss **30 cm away**	**Electrical Field** volts/metre **30 cm away**
Hair dryer	10 000-25 000	2-200*	40
Shaver	5 000-10 000	—	—
Kitchen stove	”	2	4
Food blender	1 000-5 000	—	—
Portable heater	”	10	—
Drill	”	12	—
Vacuum cleaner	10-1 000	22	—
Toaster	”	2	40
Electric blanket	10-500	2	250
Radio amplifier	—	—	90
Iron	10-100	2	60
Refrigerator	1-100	2	—
Colour TV/computer monitor	70-500	2-8 (1 at 1 m)	—

* According to the power rating of the model

ical effects of ELF radiation is the New York power lines study sponsored by the New York Power Authority. Sixteen research projects dealt with the effects of ELF electric and magnetic fields on reproduction, cell biology, neurobiology, human and animal behaviour, and the epidemiology of cancer. Among the findings were changes to the metabolism in animals' nervous systems, which affected their auditory and visual senses. All studies on animals trained in performance routines showed that ELF fields brought about behavioural changes. Human exposure resulted in a small but statistically significant change in cardiac function and a loss of ability by individuals to perform certain tasks. The research panel administering the projects concluded that: 'The variety of behavioural and nervous system effects may not constitute a major hazard because most appear to be reversible, but they may impact temporarily on human function.'[28]

A project in the New York study, the results of which were awaited with the greatest interest, was a four-year replication of the Wertheimer research by Dr David Savitz and his research team. Savitz concluded that children living in homes which had a continual exposure to a magnetic field of between two and three milligauss — roughly the amount received living 15 metres from a power line, although some lines would cause such exposure at a greater distance — had a one-and-a-half to two times greater risk of developing cancer. The Savitz team failed to find a statistically significant excess of adult cancer. Exposure to ELF magnetic fields from power lines could be responsible for 10 to 15 per cent of all childhood cancers.[29] 'The stakes are so high that society cannot sit around and do nothing waiting for another study to be done', commented Dr David Carpenter, Dean of Public Health, who headed the research panel. 'It should immediately start finding ways to minimise exposure to electromagnetic fields.'[30] Electrical utilities could reduce community exposure to the fields by further distancing power lines from houses and schools, putting them underground, or redesigning power-line configuration.[31]

UNTIL the late 1980s, utilities and radiation protection authorities in Australia took the position that there was no scientific basis for people's concern about exposure to ELF fields from power lines. The fields from the lines gave no cause for worry, they said, because they were far below the upper limit set by international safety standards. The Victorian health department's radiation committee rejected the Savitz finding, without offering concerned communities any analysis of this or other published epidemiological findings. The first serious analysis of overseas findings was carried out in 1986 by Professor Bruce Armstrong, head of the NHMRC epidemiological unit, who concluded that further investigation was warranted and that 'greatest concern should be focussed on the possibility that ELF may cause an increase in the incidence of human cancer'.[32] A long-overdue exercise by Australian radiation experts, to take a broad look at the subject of power-line ELF fields, was undertaken in 1988 at an Australian Radiation Laboratory workshop. Dr Michael Repacholi, who had headed a committee which recommended the existing international safety standard for power-line fields, told the workshop that ELF magnetic fields could possibly be a weak carcinogen.[33] In an evaluation of the cancer potential of ELF magnetic fields, Dr Vincent Delpizzo, of ARL, concluded that 'there is enough evidence at present to justify further investigation of this subject'. On the balance of probability, the findings of excess cancer are likely to find support from new research.[34]

An outcome of the Merri Creek power-line controversy was that the SECV promised to change its attitude to community consultation. According to its chairperson, Jim Smith, 'We have now established good relationships with community groups and individuals concerned about the project and we look forward to close consultation with local communities in planning future transmission lines'.[35] The winds of change had blown a little earlier in other utility board rooms. Dr Richard Phillips, of the US Environmental Protection Agency, told an international conference in 1986 that utilities 'are finally taking the ELF problem seriously. They are facing up to this issue more and more . . . Now the issue is not whether there are biological effects or possible harm but at what level and what duration of exposure'.[36]

Many more scientific papers on the health effects of ELF radiation are yet to be written. The weighing of the positive and negative findings will continue for some time. Exposed people will be the research subjects, the 'guinea pigs'. People naturally see this as evidence in itself of the possibility of harm and so of the need to keep exposure to the technically feasible minimum. And whatever the likelihood of harm, it is the young who are the most vulnerable.

Chapter 5

Electromagnetic radiation at work

> Devices and processes emitting electromagnetic radiation in Australian workplaces are largely unregulated. There are no statutory limits on the levels of EMR to which Australian workers may be exposed.
>
> — *John Mathews,* Health and Safety at Work, *p. 169, 1985.*

FOR eight years Sharon Foster worked in a small Australian plastics factory operating radio-frequency heat-sealing machines. She positioned plastic pieces between two electrodes and then pushed a button. The two electrodes closed up and microwaves passed between them, welding the pieces together. With each operation Sharon got 'the heat feeling that goes up and down your legs. You could feel heat rising up your legs and as the machine was finishing the weld you could feel the heat disappearing from your legs'.[1]

On winter mornings workers preferred machines giving the most warmth. Freezing when they arrived at the factory, they would compete for the big powerful machines, the nice warming ones. Sharon Foster was pregnant when she was working on a machine which gave more feeling of heat than most. Her pregnancy was quite normal, and Sharon felt really happy until her baby was born. Her baby had spina bifida. That made the family look back into their family history. They could find no record of the deformity.

A workmate, Marion Bacon, started to think that it must have something to do with the RF radiation. 'I wasn't unduly worried until Sharon had her baby and I started to think back about various girls who had had pregnancies', Marion Bacon recalled. 'Out of thirteen of them there were only two normal. They either miscarried or had deformed babies and that did not seem to be right. Then another girl developed leukaemia and died. You just don't have that number of things go wrong in a small factory. It just couldn't be a coincidence. There would have to be something wrong.'

When the management was told that their machines could be causing not just a heat feeling but something more serious, they simply shrugged it off derisively: 'They're only women, crazy.'[2] Only after much prodding did management have their machines tested for leaking microwaves.

A SURVEY by the Australian Radiation Laboratory of radio-frequency dielectric heaters in a number of factories has revealed well over half to be extremely leaky. At times the RF field in the air has been powerful enough to light a fluorescent tube in another room. In some factories, RF radiation around workers was shown to be hundreds of times the safe limit set by the Australian RF safety standard.[3]

The exposure of workers to EMR from heat-sealers and similar devices is largely unregulated in Australian factories. The safety standard applying to heat-sealers is voluntary, not enforceable, since there is no legislative backing for it. Thousands of RF dielectric heating devices operate in factories throughout Australia, some large and extremely powerful. Besides sealing plastics, they set glues in plywood and other laminated products, dry foodstuffs, and cook biscuits and other food.

The absorption of energy by the human body from incident RF radiation is at its peak when the radiation is emitted in the VHF band — which is the range of frequency where most industrial RF dielectric heaters in Australia operate (see box: The electromagnetic spectrum, in chapter 2). Domestic cookers use microwaves which have shorter wavelengths, more commensurate with the much smaller objects being heated. The energy of any VHF radiation absorbed by an operator of a leaky heat-sealing machine is converted to heat. But the heating effect is not uniform over the whole body; hotspots occur up to ten times the average for the whole body. Animals exposed to RF radiation have been observed to suffer adverse effects on their nervous, immune and cardiovascular systems. The worst of the leaky machines tested in Australia were found to be emitting enough RF radiation to cause temperature rises in parts of the body by four degrees Celsius. Such a temperature rise is more than enough to damage nervous tissue and the foetus of a pregnant woman.

Many people sit at RF heat-sealing machines each working day with RF waves resonating in their bodies. Thermal injury is being caused needlessly. Simple devices to shield the electrodes are available to reduce radiation emissions to within the limits set by the Standards Association of Australia (SAA). That occupational standard was set to protect against *thermal* injury, but the scientific basis for setting the standard took no account of *athermal* injury. The Australian Council of Trade Unions believes it necessary 'to study the athermal effects in detail and to include in the standard a clear notification of these effects on human health'.

When the SAA drafting committee first met, one of its members, Admiral Lloyd, proposed that consideration of athermal effects be left 'until it is necessary to study them'. Thereafter these effects were ignored in setting the standard.[4] Was this an oversight? Or does it exemplify 'a carefully orchestrated cover-up' by the military and communications industries?[5] Or is it simply a case of preferring not to know?

When Sandy Doull, a trade union health officer, drew Telecom's attention to a reported excess of leukaemia among electrical and electronic workers, they showed no willingness to investigate. 'Telecom told us,' Doull said, 'that we would have to show to them a statistically significant increase in leukaemia amongst their radio-frequency-exposed workers before they would take this problem seriously; that is leaving it far too late until you've got a group of people who are dead and can be seen to be dead.'[6]

Usually it is in a workplace that health damage from a biologically harmful agent, such as a chemical or radiation, first shows up. Someone will notice that workers handling a substance or operating particular equipment suffer some kind of sickness in common. When the possibility of an association between the work and a sickness is raised, industry management almost invariably responds by refusing to consider any chance of a connection. Where independent studies turn up positive evidence of a work-related disease, the tendency on the part of industry is to discredit the research rather than offer financial support for further investigation. Unfortunately, occupational health authorities seem to prefer to cite the negative rather than the positive evidence of an occupational health problem. The burden of proof is put on those not in a position to afford the long-term investigations that are so often required.

Even when an association between a disease and some biologically harmful agent in the workplace has been established beyond reasonable doubt, the legal hurdles to obtaining compensation can be great indeed. In the case of asbestos-related disease it took many years of legal battles in Australian courts (and union efforts in the political arena) to establish the right of victims to compensation. And this was even though inhalation of asbestos had been specifically identified as an agent which causes the fatal lung disease mesothelioma. With the present state of knowledge, an association between low levels of radiation and a cancer can be argued only on circumstantial evidence. In the case of health injury through radiation exposure, the legal battles have yet to be won in Australia. The victims, and society as a whole, would greatly benefit if ways other than lengthy, stressful and costly litigation could be devised to provide just compensation for injury from biologically active agents in the workplace environment. Society should also take all possible steps to change cavalier attitudes still prevalent in industry towards these agents.

Radar

Evidence of thermal injury due to RF radiation is admissible when claiming compensation; but evidence of athermal injury is in dispute.

In Melbourne in 1985 a legal battle for compensation for an athermal injury was fought in the obscurity of the Administrative Appeals Tribunal. Leslie Brown was claiming compensation for cataracts caused by radar.[7] He had been exposed to microwaves while maintaining radar equipment at Essendon airport between

1958 and 1965. Because a constant link needed to be kept between traffic control and flying aircraft, he carried on maintenance work while the radar was switched on. Brown had also worked on early wartime equipment known to be faulty. This meant his exposure levels were far in excess of what is now considered safe. One of the radar units he worked on is now located at a training school. It carries a warning sign: 'Do not work on the radar antenna or in its vicinity with radar TX switched on.' Brown's orders were to work while the radar was switched on. It would seem he had a very strong case.

Brown's cataracts had developed at an early age of 54 years. They were *posterior sub-capsular*, a type of cataract associated with radiation and drugs rather than the normal ageing process. Dr Markwick, Brown's operating surgeon, told the tribunal that 'his exposure to these microwaves was a factor in the production of cataracts'.[8] On the other hand, Dr Colvin, a consultant ophthalmologist to the Royal Australian Air Force, testified, 'There is absolutely no scientific evidence linking human cataracts and microwave energy.'[9]

The leading two experts to testify were Professor Fred Hollows and Colonel Budd Appleton. Hollows is an ophthalmologist at the University of New South Wales and had studied cataract formation in workers maintaining microwave transmission antennas. Appleton acts as an ophthalmology consultant to the US air force. Ironically, the evidence of both witnesses came via RF communication, the real suspect in the case. Appleton's evidence was transmitted from St Paul in the United States, and Hollows's evidence from Bourke in outback Australia:

> *Question*: You say that small repeated doses of less than 120 milliwatts per sq.cm can cause cataract? Is that correct?
>
> *Professor Hollows*: It is.
>
> *Question*: And these doses can be thermal or athermal?
>
> *Professor Hollows*: That is right.[10]
>
> *Question*: Is a connection possible?
>
> *Colonel Appleton*: If you were asking me as . . . an expert witness, which I think you are asking me in this case, I will say from the standpoint of medical likelihood there is no reason to suspect at this time any connection between microwave and cataract.[11]

The tribunal decided against Brown because it 'was unable to accept the theory propounded by Hollows. Since much of the scientific evidence emanates from the United States of America, the evidence of Dr Appleton assumes special significance'.[12] In fact, Appleton's views on radiation and health are under strong challenge in his own country.

THE most common injury presently associated with exposure to radar is the cataract, which is an opacity in the lens of the eye that prevents visible light from reaching the retina. Cataract formation from radar beams, powerful enough to cause burns, is not disputed. Athermal formation of cataracts from low-level microwaves is disputed among scientists and hotly denied by the controllers of radar.

Joe Towne, a retired US airman, decided to take the dispute out of the scientific realm and into the public arena. He had served for ten years during the cold war as a radar operator in US spy planes. In the late 1960s, after he had retired, Towne developed cataracts. He suspected radar, and his search of the scientific literature convinced him that his hunch was correct. The US Veterans Administration rebuffed all his approaches for compensation. But Towne did not give up. He set out to call public attention to his plight and that of other airmen suffering from the effects of radar. In 1971 he unearthed an official health study of the crews of the spy planes supporting his claim. The study identified not only radar but the plane's electronic equipment as a source of radiation causing cataracts.[13]

In 1976 Towne helped to form a Radar Victims Network. This brought together airmen who believed their illnesses, ranging from blindness and cancer to heart and blood disorders, to be the result of RF radiation exposure. The network was to 'help others who have suffered damage to health from their exposure to electromagnetic radiation'.[14] The network gave an information backup to the airmen, which was essential if they were to be successful in their compensation claims for radiation-related diseases.

However, to avoid setting a legal precedent, the airmen were offered out-of-court settlements on the eve of each pending trial. This is a device used to deny legal precedent to other victims making similar claims. Each claimant in turn is made to face the choice either of being bogged down in a mire of legal proceedings or of accepting an out-of-court settlement. In Australia, the victims of asbestos struggled for 30 years before they were able to establish the vital precedent that gave them the legal right to adequate compensation.

Milton Zaret, an ophthalmologist at New York University, is a central figure in the athermal cataract controversy. Zaret says, 'Chronic microwave cataract develops slowly over a period measured in years and follows repeated irradiation at athermal intensities'.[15] Cloudi ps at the rear (posterior) of the lens ract results from cloudiness inside the red posterior sub-capsular cataracts. number of scientists who, from the sive use of thermal criteria to set RF

Zaret testified on behalf of victims of at he found his scientific credibility n a successful compensation claim by th of her husband, Samuel Yannon, nsmitters on the top of the Empire . Twice a day, Yannon aligned the 14 years on the job, his previously teriorate suddenly. His hearing and nuscular co-ordination. Zaret gave F radiation contributed to Yannon's idence was admitted; and a step had been taken in the struggle for the right to compensation for *delayed* health damage that can result from occupational exposure to RF radiation.

Microwaves

The antennas on towers which transmit radio and television programs and telephone messages are a familiar sight. Linespeople climb these towers to tune and maintain the antennas. Professor Hollows studied Telecom linespeople for cataracts. He surveyed those who had been maintaining antennas for at least 15 years and who were under 60 years of age. The control group was chosen from among mail sorters.

One in every five of those exposed to microwaves was found to have developed a posterior sub-capsular cataract. The incidence was about three times that of the control group. Throughout the survey Hollows worked blind; that is, he was kept ignorant of who had been exposed to microwaves and who had not. 'The results', Hollows said in a cautious note to the British medical journal *Lancet*, 'suggest that an increase in the posterior sub-capsular cataract in radiolinesmen may be work-related'.[16]

Besides the factor of microwave exposure singled out for study, there were other factors in the working environment of the two groups that were difficult to balance. Working outdoors meant greater exposure to ultraviolet radiation. Working on the wiring meant exposure to toxic fumes from fluxes and plastics. With the variables so difficult to match, Hollows was naturally cautious in interpreting his results. But since the study showed a definite health effect, trade union health officers argued the need for caution. They insisted maintenance work be carried out only with the beam turned off.

The quality of the research has not been questioned, but the results were not gratefully received by Telecom. Hollows had not obtained Telecom's agreement to study its workers. 'Dr Hocking [Telecom's chief industrial hygienist] wrote to the vice-chancellor of my university,' Fred Hollows said, 'suggesting that it was improper for me to publish this stuff in *Lancet*, which I thought was getting a little bit heavy'.[17]

Hollows was refused a grant from the NHMRC for follow-up research, despite a recommendation by the NHMRC's own referees. A second application was also refused. Thus replication studies of considerable significance for industrial health were closed off by this funding body.

Sandy Doull believes a roadblock has developed to stop research into chronic RF radiation exposure because doing anything at all would be 'tantamount to admitting all may not be well, and no admission of that sort is permissible. It is a form of paralysis that ensures that the problem gets worse'.[18]

Meanwhile, tribunals reject athermal injury as inadmissible — as happened in Leslie Brown's case — simply because it lacks a 'credible' scientific basis. This is a catch-22 situation with potentially serious, adverse public-health implications.

The VDT

Visual display terminals (VDTs), or computer monitors, are everywhere: in schools, workplaces, homes, fun parlours, banks, ticket reservation counters, and shops. In data-processing centres, commercial houses and government departments, newspaper offices and now schools, VDTs are flooding in.

In Australia, hundreds of thousands of people sit in front of a VDT every workday. Ninety per cent are women. Children sit for hours at their videos playing games, doing lessons and solving problems. All the time low-level non-ionising EMR is being emitted by the electronic circuitry.

The ARL conducted a survey of VDTs available on the Australian market and found that in all cases RF emissions were below the safe level set by the International Radiation Protection Association (IRPA).[19] The VDTs were given a clean bill of health. Again, an official radiation standard was an unquestioned benchmark of safety. Again, an important question was not asked: what is the lowest *technically achievable* level of RF emissions?

The Swedish Institute of Radiation Protection adopts a more questioning attitude than its kindred Australian body. The Swedish body says that investigations to date do 'not give a complete picture of the total radiation environment around a VDT'. The reason is that not enough is known about the *pulsing* magnetic fields. These fields induce pulsing currents in the body. 'The eventual effect of these pulsing currents cannot be determined with the present state of knowledge.'[20]

Unlike the Australian radiation and health authorities, the Swedish body is promoting studies looking into the possible health effects of weak, pulsed magnetic fields where they might expect to find them. One study, by Dr Bernhardt Tribukait at the Karolinska institute in Stockholm in 1986, found indications of an increased number of malformations among pregnant mice exposed to the fields compared with unexposed animals. In 1987 Dr Hakon Froelen and his research team at the University of Agricultural Sciences at Uppsala found a significant increase in the number of placental resorptions (similar to miscarriages) in pregnant mice exposed to weak, pulsed magnetic fields of the kind that emanate from VDTs. A second, larger study showed the same result. In a follow-up study Froelen found the foetuses of mice to be most sensitive in the early stages of pregnancy — which is consistent with the effects of other kinds of radiation.[21]

Further evidence that fields of the nature of those emitted from VDTs were biologically active came from a recent international study on chicken embryos involving six laboratories in Sweden, Canada, Spain and the United States. The combined results of these six laboratories indicated an increase in the number of birth abnormalities.[22]

Epidemiological studies on pregnant women, working as keyboard operators, have indicated work-related adverse health outcomes. At a conference in Stockholm in 1986, excess miscarriage and cardiovascular abnormalities among offspring were reported. Another report to the conference pointed to stress rather than RF radiation as the source of birth abnormalities among the operators.[23] However, the evidence that exposure to the emissions from VDTs poses reproductive hazards for women keyboard operators is consistent with many other findings of adverse outcomes of pregnancy induced by non-ionising EMR. In 1988 a survey team, headed by Marilyn Goldhaber, provided additional evidence when they reported that pregnant women who worked at the keyboard experienced approximately 80 per cent more miscarriages than women who did similar work without VDTs. Women who worked at the keyboard less than 20 hours each week showed no statistical increase in adverse birth outcomes. After reviewing past studies similar to their own, the Goldhaber team came to the conclusion that 'consistent evidence across studies provides some basis to suspect that an excess risk could be real'. However, the team was not able to exclude entirely the possibility that job stress was a factor in the increased risk.[24] Further US research, reported in mid-March 1991, has disputed Goldhaber's key finding.

The difficulties in isolating the emissions from many other hazards in the work conditions associated with VDTs are generally acknowledged. Keyboard work is stressful, and operators often have been forced to adopt unhealthy postures because of improper ergonomic practices that are now being corrected. On the other hand, the foetus is known to be extremely sensitive to radiation. In some Eastern European countries, pregnant women are transferred out of any radiation environment in their workplace.

MOST VDTs operate in much the same way as television sets: they use a cathode-ray tube to display information. Images form on a phosphor coating as the electron beam, pulsing at around 15 000 times a second (15 000Hz), scans the screen. Electric currents in the circuitry emit pulsed radiations. As the electron beam strikes the glass, weak X-rays are produced. However, by operating below a certain voltage and using proper design principles, X-ray emissions become negligible. But 'lemons' do come off production lines. In the past, some television sets and VDTs have emitted excessive X-rays. Workers can only protect themselves against faulty equipment emitting invisible radiation by insisting on independent monitoring.

Visible light is, of course, emitted from the screen image, and eye defects may develop among susceptible operators. Traces of ultraviolet rays and microwaves are emitted. Like other electrical equipment powered by alternating current, VDTs emit ELF electric and magnetic fields. Close to the screen the magnetic field can be around 70 milligauss or even much higher. The fields are most intense at the sides and back of a monitor. The fields decay rapidly with distance and, at the keyboard (30 cm), the ELF magnetic field is generally down to around two to four milligauss. At an arm's length from the screen the field drops to below one milligauss (see box: Electromagnetic fields surrounding electrical equipment, in chapter 4).

Over 90 per cent of all RF emissions from a VDT screen are pulsed very-low-frequency radiation (VLF) in the frequency range of 14 to 250kHz. These are the emissions which have caused most concern about possible health effects on operators and others working in the vicinity. The ARL survey found VLF emissions of VDTs sold in Australia to be up to one-third the permissible limit set in the IRPA safety standard. The ARL pronounced the emissions to be 'low to very low', and concluded that 'these emissions do not pose a health problem to VDT operators'.[25] However, the difference between the lowest and highest emissions was more than a hundredfold.

As a VDT user myself, I wrote to the ARL asking that they

emphasise that some VDTs are extremely weak RF emitters. After all, safety standards say that radiation exposure should be kept as 'low as reasonably achievable'. This is the ALARA principle. 'Whilst we would certainly advocate adherence to the ALARA principle', the laboratory replied, 'it is up to others (for example, VDT purchasers), given the results of our measurements, to decide the relevance of ALARA.' However, since the emissions are only up to one-third of the permissible magnetic-field exposure, 'is it worth the extra cost?'.[26]

While operators are assured by the ARL that VLF emissions from VDTs are not worth bothering about, others take a more cautious attitude. Some manufacturers, though not all, are designing their VDTs to lower radiation emissions. A relatively inexpensive shield around the high-voltage transformer reduces emissions to a negligible level. Special shielding materials used to coat the edges of doors on microwave ovens can also be used to protect users of VDTs and other electronic equipment from RF emissions. In Sweden, unions and management have drawn up new strict safety standards, and VDTs guaranteed against RF exposure may now be purchased.

The US Office of Technological Assessment (OTA), after reviewing the research on low-frequency RF radiation, concluded that 'there are clear effects on living systems'. Although the research does not definitely demonstrate health effects, OTA says it is no longer possible to categorically assert that there are no health risks.[27]

Scientists say additional research is needed. That can only mean more epidemiological studies on pregnant women operators. Manufacturers have been most reluctant to incorporate devices that screen out the VLF waves; instead they have called for further investigation of possible health effects. 'To adopt the attitude that electronic equipment such as the VDT is innocent until proved guilty,' said investigative journalist Paul Brodeur, long an advocate of improving RF safety practices, 'should be rejected by sensible people everywhere. To do otherwise is to accept a situation in which millions of people continue to be treated as animals in long-term biological experiments whose consequences remain unknown.'[28] IBM has since announced that it plans to incorporate a device into its monitors to prevent the emission of VLF radiation. Could this have been in response to Brodeur's strong words in the reputable *New Yorker* magazine?

The primary objective of radiation protection must be to set as low an exposure-level as possible. Setting permissible exposures below an estimated danger level is starting at the wrong end. 'The assumption is made that when one is walking along the edge of a substantially dangerous cliff,' Professor Steneck said, 'the best way to keep from falling is to know exactly where the edge is. Theoretically, if all the twists and turns were properly plotted one would not fall. It is difficult not to point out in this regard that one could also avoid falling by not walking so close to the edge.'

To expand this analogy a little, David Hollway said, 'Published research points out that this cliff is probably undercut in places. The pressure to site the footpath as close to the edge of the cliff as possible comes from those who own the real estate, which is in this case the RF spectrum'.[29]

PART II

Nuclear Sources of Radiation

Alpha, beta, and gamma rays and neutrons are emitted as the *nuclei* of radioactive atoms, or *radio nuclides,* disintegrate. Gamma rays are electromagnetic radiation (EMR); the other radiations are *atomic or sub-atomic particles.* Like X-rays they are *ionising* radiations.

Radiation emissions from *non-nuclear* sources last only so long as the source is switched on. The emissions from *nuclear* sources may persist for great spans of time. The development of the nuclear reactor using uranium as the fuel opened the gates to a comparative flood of long-lived radioactive wastes.

Once radionuclides are released into the environment their radiations are beyond human control. Radionuclides enter ecosystems where they reconcentrate as they move along food chains. The advent of nuclear industry has led to increases of radioactivity in the human food supply.

The association of a high incidence of cancer with fallout from nuclear explosions and the operation of nuclear plants has led to a growing public concern about all nuclear activities.

Chapter 6

Radioactive beginnings: radium, health spas and mineral waters

> The danger signals went unheeded. Between 1915 and 1930 thousands of people in the United States actually ate and drank radium. Patients wealthy enough to afford this "cure" took radium water or injections of radium salts for all kinds of diseases.
>
> — *Jack Schubert & Ralph Lapp,* Radiation — What it is and how it affects you, *p. 112, 1957.*

THE French physicist Henri Becquerel first heard of X-rays when, in January 1896, he attended a meeting of the Académie des Sciences in Paris. His interest was aroused by the lecturer's reference to X-rays emitted from the luminescence produced on the glass wall of a cathode-ray tube when struck by the electron beam. Becquerel had studied the luminescence of uranium solids caused by sunlight. This luminescence, he thought, might also be a source of X-rays.

Back in his laboratory Becquerel put a little pile of a uranium substance on top of a black paper wrapped around a photographic plate. He exposed the uranium to sunlight. A dark shadow, the image of the pile, appeared on the photographic plate. Becquerel thought the image was caused by a penetrating radiation that must have come from the uranium's luminescence. Then he had a lucky accident. One cloudy day he decided to put his little uranium pile, sitting on top of a wrapped photographic plate, away in a closed drawer. Days later, when he looked at the plate he found, to his great surprise, the same shadow on the plate. So, even in total darkness, uranium emitted radiation. Luminescence could not be its source. A uranium substance must be constantly emitting radiant energy.

Becquerel's radiation did not create the sensation Roentgen's X-rays did. His discovery was ignored by the newspapers. Becquerel rays, as they were at first called, were not hailed as X-rays had been, as a major contribution to scientific knowledge. Colleagues evinced little curiosity about emanations of uranium. Uranium did not produce intense beams like X-ray machines, and no one could think of a use for the weak radiation. It could not, for instance, be adapted to public displays of shadow pictures on a fluoroscope. So, Becquerel soon turned his attention to more interesting fields of scientific endeavour.

It was left to one of Becquerel's students, Marie Curie, to investigate uranium and its radiation. She found that uranium ore gave off far more radiation than could possibly be explained by the concentration of uranium she had found in the ore. Curie also found that thorium emitted radiation, and proposed the term *radioactive* to describe substances emitting radiation.

Marie and her husband Pierre began an intensive search for radioactive elements. In 1898, after two years of unrelenting toil refining residues from many tonnes of a uranium ore, called pitchblende, the Curies were rewarded. First, radioactive polonium was isolated. Then came radium, far more radioactive than uranium; and in quantities which, though extremely minute, were enough to investigate.

Radium

The glow from specks of radium stirred popular imagination. They aroused newspaper speculation on possible benefits of radioactivity. The scarcity and expense of radium did not deter extravagant suggestions about its possible application. Perpetual light for bicycles at night was proposed. Radioactive beverages that 'glowed in the dark when lifted to the lips' were suggested. Luminous bodies performed 'radium dances' staged in a darkened theatre![1]

Scientists saw more serious uses for a highly radioactive source. Ernest Rutherford, the eminent nuclear physicist, and other scientists quickly realised how the rays might be used to explore the *inside* of atoms; in some fascinatingly simple experiments they used the rays to reveal the basic structure of atoms. The atom was found to have a *nucleus* (core) with *electrons* orbiting around it. Radioactivity occurred when the nuclei of radium and other radioactive atoms, known as *radionuclides*, disintegrated. The pioneers of atomic investigation soon realised that the nucleus must indeed be an immense source of energy, so the search began to discover how to release atomic energy (see appendix: Where radiations come from).

Radiations emanating from radioactive substances were soon recognised as being of three kinds: called simply *alpha*, *beta* and *gamma* rays. Two, alpha and beta, behave as streams of particles: alpha rays are positively charged helium nuclei; beta rays are negatively charged electrons; gamma rays are electromagnetic radiation like X-rays, only higher in frequency and even more energetic than X-rays. All three are *ionising*, like X-rays.

Alpha particles have little penetrating power. They are stopped at the outer layers of dead skin. Beta particles penetrate up to 20 mm inside the body. Gamma rays, like X-rays, penetrate deep into the body. When gamma rays are wanted for irradiation, the alpha and beta rays may be filtered out by holding the radium or other radionuclide in a suitable container (see Figure 6.1).

Figure 6.1: The penetrating power of alpha, beta and gamma rays

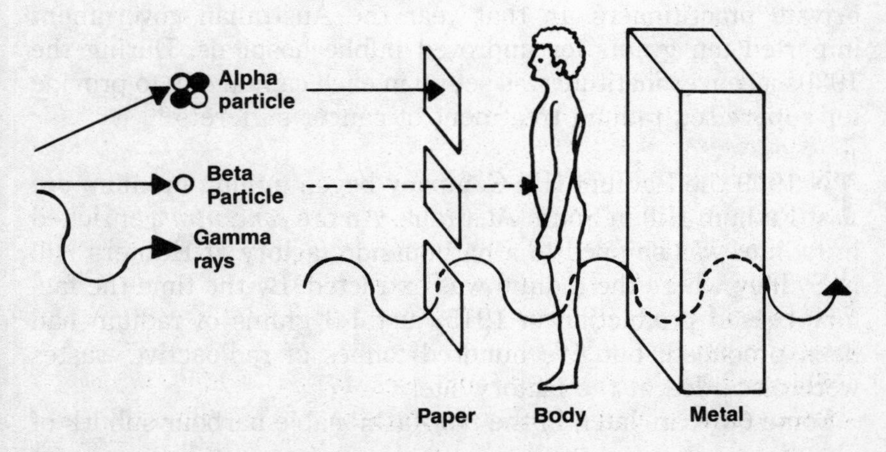

The biological activity of radium soon made itself felt. Pierre Curie carried a glass vial of radium in his pocket for just a few days; skin burns and ulceration appeared on his skin a few weeks later. The gamma rays were evidently causing the same tissue damage as X-rays. It was not long before radium was used in the new radiotherapies then being developed with X-rays. In 1903, Becquerel and the Curies were jointly awarded the Nobel Prize for Physics.

The Curies helped to set up a factory in Paris to isolate radium from pitchblende. A minute quantity of radium arrived in Australia in 1903. Its radioactivity was demonstrated during a science exposition in Sydney, where the invisible incessant emanation was revealed by a flicker on a fluorescent screen. But there was little radium around to satisfy an awakened medical interest. Herman Lawrence, a Melbourne dermatologist, told a medical congress in 1905 that 'specimens of radium — at least, of any therapeutical value — are still practically unobtainable out here in Australia'. However, he went on to say that radium, scarce though it was, had advantages over X-ray machines because 'your radium specimen is always ready for use — it cannot get out of order'.[2] Lawrence was among the first to use radium regularly in radiotherapy. He used it to treat rodent ulcers, cancer, eczema and even warts.

By 1915 the amount of radium isolated worldwide was still only about seven grams. In Australia by 1927, the radium available for medical use amounted to about a gram, mostly held by private practitioners. In that year the Australian government imported ten grams for approved public hospitals. During the 1930s a cancer institute was set up in each capital city to provide for supervised radium treatment of cancer sufferers.[3]

IN 1909 the Radium Hill Company began mining uranium ore at Radium Hill in South Australia. An ore *concentrate* enriched in radium was shipped to a harbourside factory at Hunters Hill in Sydney where the radium was extracted. By the time the factory ceased production in 1915, just 1.3 grams of radium had been produced, but five hundred tonnes of radioactive wastes were discarded at the factory site.

Some 60 years later, in the now-fashionable harbour suburb of

Hunters Hill, radiation from radioactive radon gas was detected in residences built on the old factory site. The radon came, of course, from the long-forgotten wastes left behind after the radium extraction. But where would the government find a new resting place for them? Understandably, no one wants radioactive wastes in their own backyard. Over a decade later the government had yet to find a resting place for the rediscovered wastes. Meanwhile empty radioactive houses remain boarded up. There are many versions of this story, in which radioactive wastes remain to haunt future generations. They relate as much to thorium wastes from mineral sands mining as from uranium mining.[4]

Until 1943, the largest radium production facility in the world was operating at Canonsburg in the United States. The factory wastes were sand-like and local people found them a good soil conditioner. So, in their wheelbarrows, they hauled the innocent-looking 'sand' and dug it into their gardens to lighten their clay soils. The town council built a recreation ground over a pit filled with the radioactive wastes. The US Atomic Energy Commission (USAEC) was aware of the radioactivity of the waste from a survey it did in the early 1960s. In fact, the USAEC's earlier 1940 records reveal that the waste was sufficiently high-grade in uranium to be 'mined' for use in the wartime Manhattan atomic bomb project.

'The AEC did not shout this very loud', Dr Edward Radford recalled as he was giving evidence on the hazards of low-level ionising radiation to the Royal Commission into British Nuclear Tests in Australia. 'It was not until the late seventies that the local citizens and the state realised that they had a serious radiation problem on their hands.' It has now become a big political issue. 'People took wood out of the old plant and built furniture and things and put it in their homes.' The wood was contaminated with radium. A survey showed a high rate of thyroid diseases in the area.

'I might say, though', Radford explained to the commissioners, 'that the radiation exposures to gamma rays were in the order of two to three times normal background, and that is why I was surprised.'[5] Radford is one of the few scientists who risked his reputation by speaking out early about the health risks associated with extremely low-level ionising radiation. Yet even he was surprised that such a low radiation level at Canonsburg still caused a statistically significant increase in cancer.

HOUR-MARKS on watch dials were once painted with a radium paint to make them luminous in the dark. It is repetitive precision work traditionally offered to women as a work opportunity. The women were taught to dip hair brushes into a cup of canary yellow paint containing radium and then to point the bristles by touching them on their lips. Each time they put the brush to their lips they would swallow a minute trace of radium.

The tragedy that unfolded can be demonstrated in the story of a single dial-painting workshop in New Jersey, which opened in 1915. At the end of the working day the factory manager would turn off the lights to show how much radium was being splattered on the floor and wasted. The greenish glow was a pretty sight. Some women even painted their teeth to make them glow in the dark. An epidemic of crippling bone disease broke out among the women dial painters. A local doctor became convinced that it was linked to the radium the women worked with. Examination showed that radium was being absorbed by their bones. The quantity in each individual's bones was less than a millionth of a gram. This was enough.

Radium is most damaging to cells when its short-range, but extremely powerful, alpha particles are emitted inside the body. Once radium (which behaves in the body like calcium) lodges in bone tissue, the cells become damaged and turn malignant. The result is crippling bone deterioration, and few early dial painters escaped. The fate of Australian dial painters and workers at the Hunters Hill radium factory can only be guessed at.

The level of alpha rays required to cause malignancies — learned from the illnesses of the dial painters — became the basis for safety practices on the handling of plutonium in the wartime Manhattan Project.[6]

Despite the fate of the dial painters, many patients continued to be treated by radium injections. As late as 1932, radium injection was still listed by the American Medical Association among its remedies. Radium injections were prescribed for children suffering from tuberculosis of the bone until 1951, when it became starkly obvious that radium was completely ineffective as a treatment.[7]

Today, tritium is used as the radioactive component in luminous dial paints. Tritium behaves chemically like hydrogen and so can migrate throughout the body. It emits beta particles, which are less damaging to living tissue than radium's alpha particles. Tritium takes more than 100 years to decay; so the discarded dials of watches and other instruments must be a growing burden on the environment.

It was not until the 1940s that the handling of radioactive paints was regulated in Australia. A Victorian regulation, gazetted in 1944, states, 'No person employed with radioactive materials shall place his fingers, or any painting tools . . . near his face'; and no person, who is 'addicted to the habit of biting his finger nails or sucking his fingers' shall be employed.[8] These prohibitions are a written testimony to the unwitting martyrs among dial painters who indulged in the practice of pointing radioactive brushes on their lips. Our occupational safety regulations are often founded on the personal suffering of workers. This is partly the result of setting working levels below a supposed danger level, instead of playing safe by prescribing the 'least possible' exposure.

Health spas

For centuries, aristocrats and the bourgeoisie sought cures for their maladies, real and imagined, at spas in Europe. These spas are springs of hot water believed to be endowed with health-restoring minerals. The spa at Lourdes is possibly the most famous for miraculous cures.

Early settlers in Australia continued the old European custom of drinking spa waters and bathing in them. It is part of local lore that people taking antipodean spa waters have been observed to arrive relying on walking sticks only to walk away without them. Among the most fashionable spas were those at Hepburn Springs and Daylesford in central Victoria.

Early this century radium and radon gas were detected in spa waters. It seemed perfectly logical to believe that the mysteries of radioactivity should be bound up in some way with the healing benefits from taking the waters. Those believing spa waters had mystical healing powers had their faith strengthened by knowledge of radioactive emanations. Spa centres in Australia and elsewhere began to boast of the radioactivity of their waters. The only echo of that boast today at Australian spas is in the occasional name such as Radium Avenue in Hepburn Springs. Visitors are now told only of the familiar nutrient elements such as iron and calcium.

Newspapers carried reports of radioactive solutions to kill bacteria, cure blindness, determine the sex of unborn children, and turn the skin of black people white. One suggested treatment of stomach cancer was to drink a radioactive potion so that the cancerous part would be bathed in the 'liquid sunshine'.[9] At times, dispensers of the magical healing powers of radium waxed lyrical. 'Radium has absolutely no toxic effects', one contributor reported to the journal *Radium* in 1916, 'it being accepted as harmoniously by the human system as is sunlight by the plant.'

Orthodox medical researchers were more 'objective'. They studied the benefits of radium and radon tonics in their clinics and reported the results in reputable journals. 'The value of radon is unquestionably established in chronic and sub-acute arthritis of all kinds', Dr Rowntree reported at an American Medical Association meeting in 1913. 'Acute, sub-acute and chronic joint and muscular rheumatism (so-called); in gout, sciatica, neuralgia, polyneuritis, lumbago . . . Eight hundred and thirty-seven or over 80% of the patients, were considered to have been improved by radium emanation (radon).'[10]

'I have had but little experience in the use of radium, but a good deal of experience in the effects on my wife', a Dr Robinson commented in the discussion that followed. 'Mrs Robinson suffered arthritis deformans. After being on crutches for nearly a year she was practically cured by the use of radium and the electric-light bath . . . She drank radium three times a day, combined with oxygen gas. The result was somewhat marvellous . . .'[11]

IT is a sobering thought that, by about the time the popularity of spa waters reached its peak, lung cancer in uranium miners had been identified with the breathing in of radon. Since about the sixteenth century, uranium miners had been known to die from a peculiar lung disease. To the German miners in the Erz mountains it was the dreaded 'bergkrankheit' or mountain disease. More than half of the Erz miners died from it. Since the 1880s the disease has been known to be a malignancy in the lung. It was only a matter of time before bergkrankheit was attributed to the radioactivity of the radon emanating from the radium in the ore. Radon will continue to take its toll of miners as long as uranium is mined.

Radon levels above spa water can be unhealthily high — up to ten times the normal level in the atmosphere. It may be 100 times higher in the health spa's bathrooms. Thermal galleries were built in Europe to provide radon-inhalation therapy; there, radon concentrations 30 000 times higher than in normal air could be reached. Radon therapy at spas persisted long after its only effect was known to be an increased risk of lung cancer. The milder spa therapies continue to this day.

IN the ranks of the medical profession, enthusiasts for radioactive cures reasoned that radioactivity introduced into the blood stream would exert its influence directly and be more effective. Patients were injected with radium solutions and, later, radon solutions.[12]

Thousands of these injections were administered to patients in the United States and Europe. Doctors administered radioactive injections to their families. In Australia there is no record of how the small quantity of radium in the hands of private practitioners was used. Until the late 1920s they had no more than a gram available for their 'cures'. Scarcity of the medicine in this instance no doubt helped their poor patients.

Radium injection has been prescribed for almost every conceivable ailment: hypertension, acne, heart problems, sexual impotence, ulcers, gout, arthritis, neuritis, diabetes, mental disorders, syphilis, sciatica, wound infections and skin disorders. The doses were often massive in the light of the now-known toxicity of radium.

Growing unease about the more reckless uses of radium therapy led to a Victorian government inquiry in 1910. The chairperson of the Board of Public Health, Mr Ham, conducted the inquiry and recommended that the use of radium 'should be limited to recognised experts, in view of the cost and dangerous potentialities of such substances'.[13]

Herschel Harris, the pioneer radiologist at Sydney Hospital, argued at the 1911 Australasian Medical Congress for the curbing of the 'booming of radium at the expense of X-rays'. Harris was among the more cautious of the early radiologists. What he said suggests private practitioners were overusing radium, despite its scarcity.

In the same year, Herman Lawrence published his book *Radium: When and How To Use It*. He reported there was little evidence to 'lead one to say that radium was a cure for cancer'. He had noted 'apparent cure of some growth'.[14] Lawrence also decried the building up of public expectations of miraculous cures from radium. Nonetheless Lawrence had a thriving business in a range of radium therapies at his Collins Street practice in Melbourne.

Though there was only very little radium in the hands of practitioners up to the 1920s, the potential for harm was still considerable. Pieces of radium of about 50 milligrams were used in the form of varnished 'plates'. Later, thin metal tubes or 'needles' were used. Since the radioactivity of radium changes only very slowly — it has a *half-life* of 1602 years (that is, the number of years before half the atoms have decayed) the needles could be used over and over again to treat many patients.

By the 1920s, solutions of radon gas were being distributed instead of radium. The radon gas was trapped in water as it emanated from radium. Because the radon has a half-life of only 3.8 days, it became harmless a short time after being distributed for authorised treatments; meanwhile the radium was held under strict control at a central government laboratory.

Radioactive mineral waters

'In the early 1920s', according to Sir Mark Oliphant, one of Australia's nuclear pioneers and, much later, a governor of South Australia, 'Professor Mawson had found a small spring in the Flinders Ranges, which produced radon and so was radioactive. At that time drinking slightly radioactive water was supposed to be good for the health. So Professor Kerr-Grant became interested in the possibility of exploiting the spring commercially.'[15]

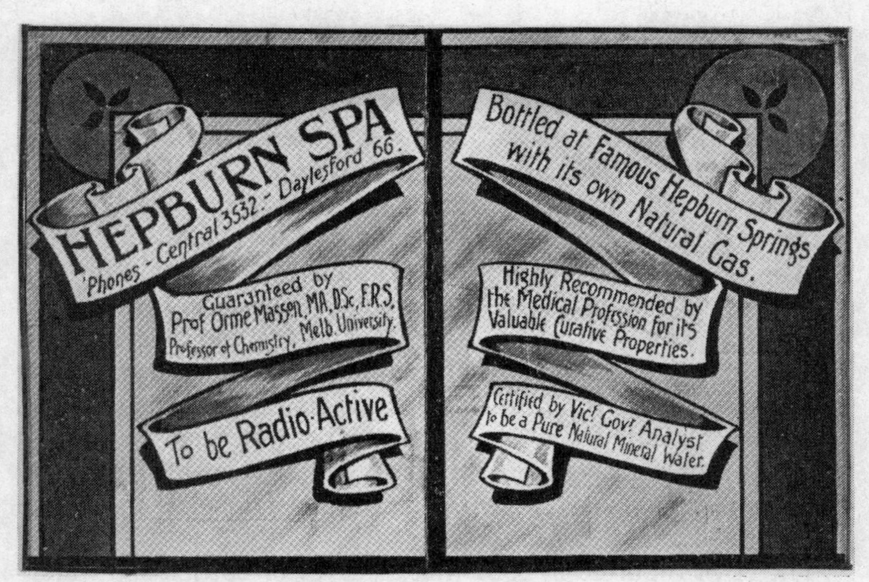

Mineral spring waters are no longer guaranteed radioactive as they once were.

Source: A collection of old prints from the *Daylesford Advocate*.

A professor promoting a radioactive tonic sounds outlandish today. But there is a touch of the entrepreneur in most of us, and tonics have been a much peddled commodity through the ages. It should make us wary of panaceas, whether they are promoted by orthodox or alternative healers.

It is hard for us to imagine seeing on shop shelves bottles of mineral water labelled 'Guaranteed Radioactive'. Yet that is how mineral waters were promoted, not so many years ago, in Australia and overseas. The inhabitants of the Brazilian town of Guarapari prospered until recently, providing for tourists who came to irradiate themselves on the local beach. But not with the sun's ultraviolet rays: the irradiating agent they sought was gamma rays, advertised as beneficial to health. The Guarapari sands, like some in Australia, contain significant quantities of monazite — a mineral made radioactive by the thorium it contains. In recent years the radioactive wastes of mineral sands have become a matter of great concern to environmentalists in Australia.

Natural mineral waters are no longer labelled radioactive; spas no longer play on the healing qualities of radon. Nonetheless, these waters are again enjoying a renewed popularity, although some are less natural than they once were. In one brand of 'natural' mineral water the radium is removed before bottling.

Today the radioactivity of ordinary drinking waters and natural mineral waters is monitored. A report by the World Health Organisation in 1979 shows that radium concentration in ordinary drinking water can be high in some places.[16] In fact, radium is the main source of the radioactivity of drinking water. It has been found that the amount of radium in people's bones is high where it is high in their drinking water. Radiation biologist John Gofman calculated that even the average radium content of drinking water in the United States could be responsible for 300 bone-cancer deaths each year.[17] This is not far below the actual numbers suffering bone cancer.

The ARL reported in 1981 that a survey of bottled spa waters sold in Australia showed that 'the levels of natural radioactivity in some cases exceed the limits set by Australian and overseas authorities for drinking water'. In fact, half of the spa waters analysed exceeded the WHO allowable level for radium content.[18] The water from Taurina Spa at Helidon in Queensland was withdrawn from sale because of its radioactivity.

The ARL report concluded that the risks to health 'are extremely small compared with other risks of everyday life'. This kind of risk comparison is unscientific and does little credit to an institution that is responsible for radiation protection. Why then bother to set a health standard? Presumably because the *collective* dose of radium to a whole population should, from a public health point of view, be kept to a minimum.

For a long time the Northern Territory health department kept quiet about the high radium content of the water supply to tourist centres and mining settlements in the Kakadu National Park. The radium level is about four times the maximum set in the WHO standard and originates from natural radioactive sources. Water is now carted to the centres and settlements.

'The risk should be neither minimised nor exaggerated', John Gofman commented, in regard to radium in drinking water. 'Rather, it should be honestly explained.'[19] Why then, we may ask, are radiation officials so ready to play down radiation risks as not worth worrying about? Why don't they honestly explain its significance for public health? It could be because so many were initiated into radiation safety practices at nuclear weapons test sites. There they rubbed shoulders with the military. Radiation protection officers trained soldiers for the radioactive battlefield. They also attempted to cultivate an acceptance of radioactive fallout as a fact of life. As we shall see in the following chapters, the performance of radiation officials at British atomic bomb test sites in Australia suggests that radiation protection got off to a bad start in this country.

Chapter 7

The casualties of atomic warfare: the Hibakusha

> Some part of the energy set free by the bomb goes to produce radioactive substances and these will emit very powerful and dangerous radiations . . . Owing to the spread of radioactive substances with the wind, the bomb could possibly not be used without killing large numbers of civilians, and this may make it unsuitable for use . . .
>
> — *Otto Frisch and Rudolf Peierls,* On the construction of a "super-bomb" based on a nuclear chain reaction in uranium, *secret memorandum to the British Government, 1940.*

> People who died and those who survived the nuclear blast in the outreaches of the two cities are called the Hibakusha — Victims of the Bombs.
>
> — *International Symposium on the Damage and After-Effects of the Bombing of Hiroshima and Nagasaki, 1977.*

AS early as 1904, the chemist Frederick Soddy, an associate of Ernest Rutherford, told a military academy in Britain that whoever controlled the atom's energy 'would possess a weapon by which he could destroy the earth'.

As war clouds gathered over Europe in 1938, reports in the science journal *Nature* sent a flurry of excitement through the international community of nuclear physicists. Uranium atoms had been split when bombarded with nuclear particles called *neutrons*. As the uranium atoms split, out shot more neutrons, which split increasing numbers of atoms in their turn. A *chain reaction* was possible! The door had been opened to the fulfilment of Soddy's fateful prophecy of an atomic bomb.

Uranium atoms exist in nature in two forms. Uranium-235, the one that splits readily, makes up less than one per cent of the whole. To make a bomb, uranium-235 has to be separated from the more common uranium-238. Such a separation was possible, though tedious for large quantities. In 1940 two physicists, Otto Frisch and Rudolf Peierls, calculated that a quite moderate quantity of uranium-235 would be enough to make an atomic bomb. The required quantity was called *critical mass*.

Later another path to the bomb opened up with the creation and discovery of plutonium. Uranium-238 does not split, but its nucleus can absorb a neutron to give plutonium-239. Plutonium splits more readily than uranium-235.

IN 1941 the Manhattan Project was set up in the United States to build atomic bombs. When Germany was defeated in 1944, no bomb had yet been assembled. Efforts were redoubled to have a bomb ready before a tottering Japan surrendered. If a bomb were not ready before then, the opportunity would be lost to test one under war conditions.

In July 1945, at Alamogordo in New Mexico, a plutonium bomb — code-named 'Trinity' — was exploded successfully. Measurements of the bomb's shock wave showed it to be equivalent to an explosion of 20 000 tonnes of TNT. A chain reaction had indeed produced an extremely powerful explosion, as scientists had predicted. The bomb's builders were somewhat terrified at the awesome power of the blast, while pleased with their achievement. They gave little thought to the radioactive cloud drifting away from the site of the explosion. The bomb's builders did not consider the fallout of radioactive debris from the cloud to be a major problem. Later, of course, radioactivity became a central issue of the nuclear debate.

A presidential committee had been appointed in the previous May to advise on the use of the bomb. The military and scientists wanted a 'virgin' target to assess fully the bomb's performance, without the complication of previous bomb damage. According to the presidential committee the most desirable target would be 'a vital war plant employing a large number of workers and closely surrounded by workers' houses'. A few Japanese cities had been preserved to provide just such a target. Hiroshima was one of those cities, and the presidential committee decided that it should be the first.[1]

An atomic attack would be different to conventional aerial bombardments, devastating as these had become with the growing ferocity of the war. Early in 1945, before the atomic bombings, an incendiary bombing raid by waves of planes on Tokyo had killed or injured 180 000 people and obliterated 50 square kilometres of the city. Now only a single plane was needed to do the job. Instead of bombs raining down over several hours, a city could be utterly devastated in a split second by the explosion of a single atom bomb.

But the atom bomb was historic for more reasons than the sheer magnitude of its blast and the immediate devastation it inflicted: it caused a fundamental change in the nature of war. As two of the more perceptive of the bomb's builders, Otto Frisch and Rudolf Peierls, had warned, radioactive poison would spread far beyond the blast, killing large numbers of civilians.[2] It is because of the bomb's radiation that Hiroshima and Nagasaki are remembered; the mass fire-bombings of Tokyo and other cities, which were more physically damaging at the time, have merged in people's memories with the widespread destruction of World War II.

On 6 August 1945 an atom bomb, 'Little Boy', fuelled with uranium-235, was exploded over Hiroshima. A plutonium bomb, 'Fat Man', was exploded over Nagasaki three days later, when Japan was already on its knees.

WHEN news of the atomic bombing of Hiroshima was broken to the world, the radiation peril was eclipsed for a time by the awesome destructive power of just one bomb. Vast fields of rubble stretched for kilometres. Human bodies were vaporised to charred sculptures or mere shadows on stone by the fierce heat of the blast. Firestorms consumed the ruins, and hot whirlwinds spindled out from the devastated city.

As the clouds of smoke swept downwind from the conflagration they dropped a muddy 'black rain', which deposited sticky and oily black spots. People thought it was oil raining down on them. But unlike oil, it burned and blistered people's skin and smarted their eyes. The river ran black as though it had had ink dropped into it. Poisoned fish floated on top of the water. Cattle ate grass spotted with the black raindrops and suffered diarrhoea. People beyond the blast and its fierce heat also suffered diarrhoea and nausea.[3]

Survivors tell of hearing pitiful cries coming from the dying in the rubble of Hiroshima and Nagasaki. They begged for water. A burning thirst tormented the dying; giving them water only hastened them to a merciful death. The piercing radiation had destroyed the lining of people's guts so that they leaked like a sieve. In the following weeks and months people most exposed to the intense radiation died slowly and agonisingly from haemorrhages and starvation as the food they ate passed to the bowel without its nourishment being absorbed.

Many who thought that they and their loved ones had escaped the bomb on that tragic morning were to discover, not long after, that some kind of sinister agent had entered the body and was destroying it.

'My daughter', recalled a mother harrowed by her memory of what followed the bombing, 'she had no burns and only minor external wounds. She was quite all right for a while. But on the 4th of September [four weeks after the atomic explosion], she suddenly became sick. She had spots all over her body. Her hair began to fall out. She vomited clumps of blood many times . . . I felt this was a strange and horrible disease. We were all afraid of it. Even the doctor did not know what it was.'[4] After ten days of agony the daughter died.

People who died and those who survived the nuclear blast in the outreaches of the two cities are called the Hibakusha, meaning the 'victims of the bombs'. The number of deaths kept rising. By the end of 1945, 120 000 had died in Hiroshima and 70 000 in Nagasaki. By 1950 the death toll in the two cities had reached 300 000. Leukaemia cases began to appear after two years, and continued to rise until 1954 before declining. In 1955 the incidence of lung, stomach and other cancers began to rise; these malignancies continue to afflict people up to the present day.[5]

Some who were in their mothers' wombs at the time of the

blast suffered microcephaly, a condition with symptoms of reduced head-size and mental retardation. The primary effect is retardation of brain growth. The head grows only in response to brain size. Radiation exposure too low to cause a measurably smaller head could nevertheless deprive a child of some capacity to learn.[6]

The atomic blasts not only destroyed the cities but much of the social fabric. In the rebuilt cities the survivors among the Hibakusha found difficulty obtaining work. They were often rejected by those they sought to marry for fear they carried mutations. Rather than being consoled with special care and attention, the Hibakusha were often shunned. They were living reminders of a terrible tragedy. Most people wanted to forget.

SOME months after the surrender of Japan a contingent of the Royal Australian Air Force flew into Japan to join the allied occupation forces.

Jack Sennett recalled how he and his companions in the contingent first saw the flattened rubble, once the city of Hiroshima. To the war-conditioned young men it looked at first sight rather normal, not unlike other fire-bombed Japanese cities they had just flown over. Incendiary bombing had levelled many cities to a plain of ashes and rubble. They got a closer look as the plane banked over the Ota River. 'We looked again', recalled Jack Sennett, 'and for the first time felt a chill about the victory won in this "just" war. As we circled over the dead city on that calm evening the brain tried to absorb a simple fact — one bomb did this.'

The contingent set up a radar communications station in Hiroshima. Off-duty Australians wandered along streets running through the ruins. At scattered points in the blast area they would come upon house-sized mounds of twisted melted bicycles, grim monuments to the pedalling throng of students and others riding to school and work on that fateful morning.

Japanese workers were employed at the radar station. One day a young messroom employee failed to appear. Because he had been listless he was not much missed. Later they learned he had died. He had seemed healthy enough. Enquiry among other workers brought an evasive reply — 'pica-don', meaning flashbang. The appearance of laziness had in fact been an overpowering weakness, stemming from irradiation, which crept upon him over months as his body tissues degenerated.

In defiance of orders not to fraternise with the enemy, the Australians became friendly with workers at the base. One of the Australians, Jack Sennett, said that 'radiation sickness became evident in one of my friendly mechanics'. Sennett had often visited him at home. 'He stopped work and several months later he too was dead. Many times an enquiry about someone, not seen recently, brought a quiet sad reply of pika-don.'[7]

Atomic Bomb Casualty Commission

Japan was occupied by the allied forces a month after the atomic bombings. Scientists, eager to learn about the effects of their bombs, swarmed over the ruins.

The bombs had exploded high above the cities. The heights could be calculated from fixed shadowy images of objects cast by the intense heat rays. The bomb over Hiroshima exploded at a height of 600 metres; the one over Nagasaki, 500 metres. Charred and melted materials directly below the explosion centres indicated temperatures of thousands of degrees Celsius.

Western journalists first learned about the extreme devastation of the Japanese cities from public relations briefings on the USS Missouri anchored in Tokyo Bay. Military spokespeople described the powerfully destructive effects of a nuclear explosion, but uttered no word about the radioactivity.

Instead of listening to and reporting the official version, Australian correspondent Wilfred Burchett set out on a hazardous journey to see the ruins of Hiroshima for himself. Burchett inspected the city's two remaining hospitals and became the first correspondent to tell the world about the new 'atomic plague'. Hospital head, Dr Katsubi, told Burchett, 'We don't know what we're dealing with. People come in. The symptoms are dizziness, haemorrhage, diarrhoea and later there are spots and before death the hair falls out and there is bleeding from the nose.'[8]

By presidential decree, all reports of the Atomic Bomb Casualty Commission (ABCC) were forwarded to the US departments of war and navy. Japanese working among the injured were forbidden to speak publicly about the suffering of radiation victims. It is believed that the army seized all evidence of radiation recorded by Japanese scientists, including photographic film that would indicate radiation intensity by the way it had fogged.

The army forbade publication of medical reports on radiation sickness. The media were mostly compliant. The *New York Times* reported a military denial of Burchett's account of the atomic plague. 'No Radioactivity in Hiroshima Ruins', the paper's headline ran; while its correspondent, W L Laurence, reported solely on 'the might of the bomb'; its victims interested him only as proof of that might.[9,10]

Any mention of *genbaku* — the bomb — was censored out of Japanese publications. References to the bomb and its radiation were blotted out of Japanese pleas to the outside world for aid in restoring their shattered lives. A small book describing bomb victims' sufferings, published in Japan in 1950, was suppressed by the occupation forces. A Japanese film, 'The effects of the atomic bombs on Hiroshima and Nagasaki', showing the great human suffering in the wake of the blast, was confiscated. Not until 1968 was the film released publicly.

Not all could be suppressed. Before the curtain came down, John Hersey published his compassionate story 'Hiroshima' in the *New Yorker* on 31 August 1946, revealing to the world the bomb victims' plight. The truth about the inhumanity of the new and terrible weapon began to spread. Only in the late 1950s did the occupation army adopt a more compassionate approach towards the Hibakusha. Their sufferings had been submerged under the military's single-minded drive to use them for data on the effects of radiation on human beings.

UNLIKE the powerful shock wave and intense heat, the radiation had left no visible record on the ruins; the only evidence was the radiation sickness and later the delayed cancers suffered by the Hibakusha.

In the days following the blasts, Japanese doctors had gone into the ruined cities and witnessed a mysterious wasting sickness. As they tended the sick and dying they came to realise that the symptoms they were diagnosing were like those from overexposure to X-rays and radium. The bombs must have given off radiation. Then they understood: even those dying in the first days from burns and shock were in any case doomed by intense radiation from the blast.

In the wake of the blast, those most exposed had received the greatest radiation exposure of the whole body ever experienced by humankind. Most died from radiation sickness. Those who survived beyond two weeks lost their hair. After about a month their white-blood-cell count dropped sharply; many developed bleeding gums and purpura (red and purple spots on the skin). As burns healed, deep rubbery scars, known as keloid tumours, formed. Some died of normally curable ailments, because the radiation had undermined their immune systems (see Table 7.1).

Several hundred thousand Hibakusha beyond the zone of intense radiation received low doses of radiation. The scientific corps of the occupying forces made these survivors the subjects of a vast study. Their medical records now provide the most comprehensive data available on the medical effects of low-level ionising radiation on the human body.

Because the bombs had exploded high over each city most

Table 7.1: Symptoms of early radiation sickness syndrome

Dose (milliSv)*	**Body damage**
250	Reduction in lymphocyte count.
1 000	Possible nausea, vomiting
4 000	Nausea, diarrhoea, drop in blood cell count; about 50% of exposed group die (LD 50) within weeks from failure of blood-forming organs.
5 000	Loss of hair; 50% die
6 000	Damage to stomach and intestine walls with loss of fluids; immediate radiation sickness; bloody diarrhoea, tormenting thirst; purpura (red and purple spots on skin); death within three weeks
10 000	Severe damage to central nervous system; death within days.

* See box: Ionising radiation doses to individuals and populations, in chapter 13.

radiation-exposure came directly from the explosion, not from the inhabitants swallowing or breathing radioactive dust. This meant several hundred thousand people had absorbed, under comparable conditions, varying low-level doses of radiation. The size of the dose depended on the distance from the explosions, how people were shielded by buildings, and even on whether they were lying in bed or standing.

Epidemiologists and radiation biologists could have no better offering of human subjects on which to do research. The deed had been done; ethical scruples about using humans were not in question. And since the military wanted the answers there was no shortage of funds. Only the more recent Chernobyl reactor disaster in the Ukraine has provided as large a group of irradiated people for study.

In 1947 the ABCC selected 120 000 survivors for its Life Span Study to make a detailed study of prospective medical records. The study was designed to answer many questions about the medical effects of low-level radiation. A control group of people too far away to be irradiated were included. Survivors at the time of the explosions were divided into categories according to age, sex and genetic differences. The most difficult problem was how to assess the actual dose received, not only by the whole body, but also by the organ in which a cancer developed. Survivors were grouped according to how their whole body dose fell into certain ranges — say 10 to 90 milliSv, 100 to 490 milliSv and 500 to 1000 milliSv (see box: Ionising radiation doses to individuals and populations, in chapter 13).

Scientists have put an enormous effort into calculating the radiation doses. Japanese-style houses were placed on nuclear test sites in the Nevada desert to determine how much radiation penetrated the roofs and walls. After atmospheric tests were banned under the 1963 partial test ban treaty, scientists placed an atomic reactor on top of a tower to determine levels of radiation falling on the human body outside and inside buildings.

IN the years since, the Life Span Study has recorded deaths among Hibakusha from cancer and other causes. The number of deaths from cancer has been plotted against the radiation dose. Medical records have been updated every few years. By 1985 about 40 per cent of the study group had died.

The estimates of radiation risk made by the International Commission for Radiological Protection (ICRP) are based largely on the data coming from the study. In 1977 the ICRP estimated the risk of fatal cancer to be 12.5 in a population of 100 000 exposed to 10 milliSv. Not all scientists agreed with the estimate or the almost exclusive use of the Hiroshima study data to estimate risk. A number of scientists looked at evidence gathered from nuclear industry and the medical use of X-rays. Here the dosage can be received over a period of time. Some of these studies indicated the ICRP risk-estimates were too low. The ICRP remained unmoved.

Then in 1980 the basis for calculating dosages at Hiroshima and Nagasaki was found to be fundamentally flawed. Two scientists working on nuclear weapons at the Lawrence Livermore national laboratory in California found the explosions must have yielded far fewer neutrons than previously thought — the moisture in the air over the Japanese cities would have mopped up most of the neutrons before they could irradiate the people in the study group. In the Nevada desert, where scientists had gathered a lot of their data, the air was drier and the neutrons travelled further. This meant that the cancers had been caused by a lower dose of radiation than previously believed. Therefore, a particular radiation dose carried a higher risk of cancer than had been previously calculated.

Another influence on the trend of results from the study has been the longer follow-up period. Many more survivors, who were young at the time of the explosions, are now included. Young people are much more vulnerable than adults to the induction of cancer by radiation. Forty-five years after the explosion they are adding more cancers to the total.[11] The revision of the study data has refuelled the debate over the ICRP's estimates of radiation risk. The questions raised in the debate are discussed later in this book.

From what they learned at Hiroshima and Nagasaki, the military compiled manuals on the estimated fighting capacity of soldiers exposed to radiation on future nuclear battlefields.

According to one training manual, 'a soldier exposed to 650 rads [6 500 milliGy] initially shows no symptoms but loses his effectiveness [to fight] in about two hours and can be expected to die in a few weeks under battlefield conditions'.

The Hibakusha were the first human guinea pigs used in nuclear experiments. They were not to be the last. The military and the scientists turned to testing their own soldiers for their physical and psychological endurance on simulated radioactive battlefields. The genius that had unravelled the secrets of the atom was corrupted by its unleashed power to serve the ends of military terror and destruction.

Chapter 8

The atomic soldiers of Maralinga

> *Question*: Was it in your mind that it was essential that people be made properly aware of the risks of contact with radioactivity?
>
> *Answer*: Contact with radioactivity, yes. I certainly would not have — well I would have been apprehensive if an hour after the shot somebody had gone walking blindly into the area, certainly I would.
>
> *Question*: Did you see it as one of your responsibilities as Scientific Director to make sure that the soldiers, who were not scientists, were properly informed?
>
> *Answer*: Well I think that was the military's responsibility. They were there to teach their people specifically about nuclear war.
>
> — *Lord Penney questioned at the Royal Commission into British Nuclear Tests in Australia; Transcript, p. 4429, 1985.*

BEFORE the rubble of the Second World War had been cleared away, national leaders began to build nuclear arsenals and train their atomic soldiers. Within weeks of the atomic bombings of Japan, the United States had designated sites to test its nuclear weapons. The first test explosion occurred in 1946 on Bikini Atoll, a US protectorate in the Pacific. In 1949 the Soviet Union exploded an atomic bomb in the desert lands of its minority races, in Kazakstan. By the early 1950s other industrial nations had joined the nuclear arms race. They too searched for test sites remote from their population centres.

In 1952 the Australian government agreed enthusiastically to British nuclear testing in its remote coastal and arid inland areas

— remote, that is, for most white Australians. The first test explosion occurred at the Monte Bello Islands in 1952. The tests came to an end at Maralinga in 1962 with minor trials on 'accidents' to nuclear weapons.

Nuclear scientists worked at the test sites along with the army. Lord Penney, the scientist in charge, told the Royal Commission into British Nuclear Tests in Australia that the army's purpose was 'to teach their people specifically about nuclear war'. Test sites were used as mock battlefields. Troops were ordered into the 'frontlines' close to nuclear fireballs, as part of trials devised to test the stamina of the new atomic soldier and his capacity to make equipment contaminated by radioactivity operational.

Around test sites men washed and scraped radioactive dust and grime off vehicles. At air force bases they steam-hosed fuselages contaminated by flights through mushroom clouds. A ship and its whole human contingent were decontaminated after sailing within three hours of the explosion into the path of a cloud. Elsewhere in the Pacific, US sailors scrubbed above and below deck on radioactive 'enemy' ships until they became nauseous with the heat and radiation. In one test at Maralinga men were asked to 'crawl, lie, walk and run amongst radioactive dust'.[1] The army wanted to launder contaminated uniforms to find out how to remove the radioactive dust. The scientists could only think of hoses and scrubbing brushes to remove the radioactive muck. The decontamination exercises were not only pathetic: they led to tragedy in later years in the lives of the men and their families. Quartermasters' stores today, logically, should be stocked with piles of hose coils and brushes awaiting nuclear war. Now, years after the Chernobyl reactor catastrophe, Ukrainian people are still hosing and scrubbing radioactive fallout from their houses and streets, just as men did at Maralinga 30 years earlier.

From the grand designs of ivory-tower science to the humble task of scrubbing — the outcome was as futile as it was tragic.

Maralinga

Rick Johnstone enlisted in the RAAF as a motor mechanic. He had served only a short time when he was ordered to prepare himself for a secret mission. He hoped for an overseas posting.

Instead he was drafted to a desolate construction camp in the dry scrublands of South Australia.

The place was Maralinga, aboriginal for 'field of thunder'. It was the third Australian site chosen to test British nuclear weapons. The earlier sites were the Monte Bello Islands off the west coast and Emu Fields in South Australia. Government leaders and media editors heralded Maralinga as a symbol of national achievement, Australia's entry into the nuclear age. It is now seen as a more dubious episode in the country's history, when Aboriginal land was desecrated with long-lasting radioactivity and men suffered needlessly from radiation. The search for national security under a nuclear umbrella has proved a military deception.

Johnstone was drafted into a Canadian unit detailed to salvage radioactive vehicles and decontaminate them. The aim of the exercise was to determine battleworthiness of vehicles and equipment for future nuclear wars. Importantly, the army wanted to know about the capacity of men to do the job. After six months of sweat and toil, Roadside campsite was bulldozed out of the desert. A low, commonplace steel tower stood erect amidst the scrublands. It gave no hint of its part in the holocaust to come.

At his first bomb firing, Johnstone and others turned their backs at zero minus ten seconds. As the count reached zero he felt as though a blinding light had passed right through him. He turned and saw a brilliant fireball billowing up from the ground. All this happened in eerie silence. Time seemed to pause until a shattering boom and an oven-hot wind broke the spell over the still of the early-morning desert. An enveloping sound passed by and into the scrublands. It pained the eardrums. The group watched while the mushroom cloud, sitting on top of a wispy stem, climbed skyward and then drifted away on the south-west wind.

'There were a lot of ashen faces when it stopped', Johnstone told the Royal Commission, 'even among the top brass who supposedly knew what was coming and what to expect'. His group of watchers were 11 km from ground zero. Where the bomb had been perched on its tower was a crater lined with green glass.[2]

Johnstone was given the 'drum' by two Canadian NCOs on how to decontaminate vehicles as well as himself. He began each day in a decontamination hut. Through the centre of the hut ran a partition with a stool perched on top. The partition divided the 'clean' area from the 'dirty' area. 'You took all your clothes off on the clean side', Johnstone told the Commission, 'sat your backside on that stool top, swung your legs over that and put your feet down into your clean protective clothing, your light socks, then into your boots'.[3]

He and the two NCOs would head off across the desert in a Landrover to look at vehicles. Johnstone was told where to drive: 'Head over this way, mate' or, when the Geiger counter began clicking too rapidly for comfort, 'too hot, let's get out of here'. The desert heat was trying. Dust seeped into their masks. Sweat ran down their chins, forcing them to drop their masks. They drained the sweat and donned them again. Or else they abandoned them altogether.

Outside the decontamination hut, men used compressed air to blast away at the dust clinging to their uniforms. This dust filled the air. Inside the hut they stripped, threw their clothes into a bin and showered. Then they were monitored by 'fruit machines'. The flashing lights of these machines recalled the pokies back home in the club, but now the blinking of the lights meant not the jackpot but a high radiation count. That meant more scrubbing of already red-raw skin with a nailbrush until the 'fruit machine' gave the all-clear. Only then was each man allowed to climb with relief over the partition into the 'clean' area.

Salvaged vehicles were decontaminated with a steam jet. The regulation gloves were bulky, making it difficult to hold the hose nozzle; it was easier to take them off and hold the steam hose with a sweat rag. The face mask misted, so that too was taken off. Johnstone asked the commissioners to imagine a vehicle that had just come in from the desert. The 'dif' would be caked with oily mud. When a jet of high-pressure steam hit the mud it went all over the place. Scientists and officers came to see how the decontamination was going. They did nothing about safety.

The high command acted on the old army dictum that the men did not need to learn about radiation. So radiation became something you could do nothing about; it was shrugged off or made the subject of macho jokes. The officers were little better informed. They knew no more than having to keep their men out of the 'hottest' areas.

Johnstone began having attacks of nausea and giddiness: 'I would have to stop the vehicle and walk around it to regain my composure, I reported these strange attacks to my immediate superior. We discussed them and decided they were just mild attacks of nerves.'[4]

RICK Johnstone was posted back to base near Sydney. He spent some time in hospital, where they did not seem to want to know about his sickness at Maralinga. On the way to another posting he became ill again and was admitted to a military hospital. Johnstone told the medical staff about his sickness at Maralinga and was ordered 'never again to mention the atomic tests at Maralinga'. He had signed articles of secrecy; if he spoke about what happened there he could get a gaol term. He was prescribed shock therapy. The 'effect was to blank out parts of my memory'.[5] Soon after, he was discharged on grounds of bad health. He had been classified A1 when he had signed on only a few years earlier.

When Johnstone told a therapy group about his Maralinga experiences years later, a doctor told him that 'there had never been any atomic bombs in Australia'. For a long time after that he did not mention Maralinga. Then in 1969 an ABC documentary was broadcast about the British nuclear tests in Australia. 'The doctor who said it was crap came to see me and apologised, and at last I felt I was looked upon as a person with a genuine problem.' After a long battle through the courts, Johnstone was awarded compensation for his radiation injuries. His condition is probably due to degeneration of nerve and gland tissue caused by radiation.[6]

COLIN Bird never sighted a mushroom cloud. He had witnessed nuclear destruction. Stationed in Japan in 1946, he was a member of a contingent taken to survey the Hiroshima ruins. The military used these excursions to impress on soldiers the immensity of the bomb's destructive power. Colin Bird was not informed about radiation health effects. That was taboo by order of the army. Had Bird been fully instructed in the health effects he would have worried earlier than he did about his later work.

As Lincoln bombers landed at Broome from their secret missions, Bird unbolted canisters from the wings of the planes. It was an open secret that the canisters contained radioactive material collected as the planes flew through radioactive clouds drifting away from Monte Bello. He placed the canisters in vehicles under the command of British scientists dressed in 'moon suits'.

Ground staff and aircrew on the job wore regulation dress or, in the heat of the day, much less. The scientists slipped their 'moon suits' off as they entered the room where they worked. The procedure naturally aroused suspicions in the men's minds about radiation hazards. To reassure the men an officer explained at parade that scientists only wore protective clothing because their working room had to be kept dust-free.

Weeks later, Colin Bird returned in a Lincoln bomber to his home base at Amberley in Queensland. Here the plane was joined by four others. After much indecision the planes were quarantined in a restricted zone. Ground staff were ordered to decontaminate the planes with steam hoses. Standing in overalls beneath a plane they directed the steam jet upwards to the cladding of the plane's fuselage. The holder of the hose got drenched as the condensed steam turned into spray. At each 'smoko' and lunchbreak overalls were discarded.

'I would shower using a scrubbing brush and soap, paying particular attention to fingernails, hair and crutch', Bird told the Royal Commission. 'I was never measured before having my first shower.'[7] Up to four showers were needed before a low-enough reading was reached on the Geiger counter. The routine went on for weeks. Officers offered assurances 'that we would be OK'.

Eventually the futility of the exercise could be ignored no longer. The planes were broken up and buried. It was not the end of the story for Colin Bird. He contracted cancer.

LANCE Edwards and his crew flew through radioactive clouds collecting dust samples. The crew were not issued with protective clothing. At pre-flight briefings they were ordered to bail out — that is, to fly out of the cloud — if the count got too high. During flights the radiation counter 'went off the clock'. Their plane became so radioactive it made no difference whether they were in or out of the cloud: the needle remained off-scale either way. 'We were getting massive doses of radiation and we were in the plane for three or four hours. We flew the plane for another two weeks.'[8]

Six years later Lance Edwards contracted thyroid cancer. Radioactive iodine, present in the cloud, can cause thyroid cancer.

THE army was ordered 'to discover the detailed effects of various types of explosion on equipment, stores and men with, and without, various types of protection'.[9] Lord Penney explained to the Royal Commission that the order meant exposing dummies, not men. Johnstone, Bird, Edwards and others have good reason to think otherwise: dummies were used, but so were men. Lord Penney also told the Commission that the exposure of personnel had been in accordance with a radiation code based on the existing international recommendations on safety. Radiation practices at test sites were according to the 'best advice available at the time'.

'In today's terms', Lord Penney said, 'it is not nice; it is not too dreadful'.[10] The Royal Commission found that there had been serious departures from the radiation standards recommended at the time.

Reduction in permissible radiation exposure was being seriously considered at the time of the tests. In 1957 the recommended maximum radiation dose was reduced drastically. Long before the stricter limits were finally set, safety standards were known to be based on calculations that greatly underestimated risk. By 1950 it was known that Hiroshima survivors, who had received low doses of radiation, showed a high incidence of leukaemia. Why then the hesitation in enforcing stricter radiation standards at the test sites? Could it be so as not to jeopardise the tests?

It seems that those in the high command were aware of the dangers. A secret fleet order issued by the British Admiralty in 1951 stated, 'There is one overriding principle, which must not be lost sight of. It is this: all radiation dosages, however small, are harmful. The only excuse for exposing men to it is demonstrable operational necessity'.[11]

The thinking of the army when they planned training courses for their new army of atomic soldiers was, of course, that exposure of men to radiation was necessary in the interests of national security. Such too was the thinking of some scientific leaders. Lord Cockcroft, head of the Atomic Research Establishment, said that, without a nuclear force, 'we shall sink to a second-class nation'. The nuclear experience had taught scientists to speak like generals.

MARALINGA, according to a government press release, 'is as good as a factory or laboratory working on atomic energy, and the staff are quite happy in the efficient organisation, which so rigidly enforces control of exposure to radiation'.

The first test firing at Maralinga took place in September 1956. Just before the firing the responsible federal minister, Mr Beale, gave newspaper editors a guided tour of the testing range. He explained to them that the men were to be indoctrinated in how to fight in nuclear wars. He also took the opportunity to call attention to 'alarmist' reports about radioactive fallout over Australia. He was referring to newspaper accounts of radioactive fallout from the Monte Bello tests earlier in the year. Mr Beale suggested that publishing these reports had not been in Australia's best interests. He was pleased to be able to report to a cabinet meeting that the editors had undertaken 'to minimise public alarm as much as they could and to print the facts'.[12] The editors were as good as their word. The Royal Commission found the media simply published government handouts. 'Only when things appeared to be going wrong was more information provided.'[13]

Newspapers portrayed Maralinga tests as science fiction. Maralinga was sanitised and dressed up for public consumption. Scientists were firmly in control of a complex hazardous operation with its hundreds of participants. Operations proceeded in clockwork fashion in the field and laboratories. Each soldier and technician followed detailed orders and executed them precisely. Groups entering radioactive zones worked under close supervision of vigilant radiation safety-officers. Film badges decorated their atomic-age radiation-proof suits. Afterwards the radiation dose measured by each individual's badge was carefully recorded.

Truth has often been a casualty in nuclear enterprises, military or civilian. The atomic-age clothing did not stop gamma rays. The clothing would need to be made of more than cloth. The 'goon-suits' issued to the ranks were suffocating, and they did not even stop the radioactive dust getting in. Personnel were not trained for their job, and the army's 'need to know' rule was enforced blindly, even among the technical staff. Health physics officers were not informed about the importance of radiation records. One described how, in the laboratory, they 'faked the results' by simply recording radiation levels they thought individuals had experienced.[14] Dosimeter badges were often thrown into a common bin or overboard from ships.

In fact, radiation monitoring was not intended to ensure individual safety. When it was done it was to study bomb performance and to obtain data for improving radiation measuring-devices. No serious effort was made to record the doses received by each and every individual. The Royal Commission concluded that because the record of individual doses was so scanty 'there is now little prospect of carrying out any worthwhile epidemiological study' of the personnel at the test sites.[15]

'They discussed in England for months, maybe years, that they were going to indoctrinate people — they called them indoctrinees — to see these things,' Lord Penney told the Commission. 'It had been discussed with our people [the scientists]. They [the military] had been studying nuclear war for years.'[16]

Professor Ernest Titterton headed the weapons safety committee, the Australian watchdog monitoring the tests. He explained publicly how the Maralinga tests would benefit Australia's own national security. Once Australia had its own nuclear armed forces it would be in a strong position to defend its long stretch of coastline. An invading force could be repelled 'with a force equipped with tactical A-weapons; [enemy] beachheads could be smashed wherever they occurred and troops could be brought to the areas concerned for clean-up [decontamination] operations'.[17] Presumably they would be armed with scrubbing brushes, buckets and hoses.

Defence tacticians and their scientific advisers were not the only ones thinking about nuclear war. By the 1950s hundreds of millions of people around the world had petitioned the United Nations to 'ban the bomb'. Australians had become more aware of the dangers of radioactive fallout. Public opinion surveys showed a growing number of Australians opposed to the Maralinga tests.

THE army could not afford to have their atomic soldiers infected with a fear of radioactivity. A Commonwealth Indoctrination Force was formed just before the first Maralinga test series. It was only small, comprising 250 British, Canadian, New Zealand and Australian officers. They were to be the disciples who would go back and teach other soldiers about fighting in a nuclear war.

'Indoctrinees will be exposed at a safe distance from the flash, thermal and blast effects of a nuclear explosion', according to the manual *Indoctrinee Force Instruction No. 1*. 'They will make a conducted tour of the firing area and of various items of service equipment, vehicles, structures, etc. exposed for trial purposes, both before and after firing.'[18] Groups of indoctrinees were stationed in covered trenches or tanks 1 600 metres from ground zero. Others went through the 'backs to the blast' routine further away, before advancing into the radioactive zone.

Dr John Moroney, later in charge of the health section at the Australian Radiation Laboratory, took part in the indoctrination program at Maralinga. He explained how the indoctrination was 'designed to overcome the initial terror' of nuclear blasts. After a blast, safety officers checked radiation levels before taking indoctrinees on a conducted tour to show them 'they could enter such an area without their balls dropping off'.[19]

Indoctrinee officers went back to their home units 'to pass on their experience to members of the Armed Forces at the conclusion of the trials'.[20] Learning about the chronic effects of low-level radiation, such as cancer and genetic defects, was never part of the indoctrination.

In any medical sense, indoctrinees or other soldiers at test sites were never used as 'guinea pigs'. Medical investigations were limited to urine analysis for radionuclides that might have been 'fortuitously' inhaled and, of course, the study of decontamination by showering and brushing. Soldiers were not exposed to high-level radiation. The purpose of indoctrination was to create a new breed of soldier fearless in the face of the nuclear fireball and sinister radioactivity around him.

Nuclear veterans v the experts

It is likely that the official facade over events at test sites would never have been broken through had the veterans not made a determined struggle to have their case heard for compensation for radiation injury.

By the late 1970s the credibility of official and media accounts of safety at the test sites was being seriously questioned. Veterans were at last telling their side of the story publicly. In the United States and Britain, veterans were fighting the same compensation cause.[21] In desperation, technicians at the French Mururoa test sites broke their pledge of secrecy to tell about the fear they held for their safety from radiation exposures.[22] Nothing is known about the thinking of Chinese personnel at test sites in the Sinkiang desert or about soldiers at Russian sites in the Kazakstan desert.

For a long time, veterans' appeals for compensation failed to shift the Australian government. 'The government is satisfied on the basis of reports submitted at the time', the minister responsible, senator Carrick, said, 'that all personnel working at Maralinga were subject to stringent health procedures'.[23]

When the Labor Party came to office in 1983 it equivocated on the veterans' claims but then gave a little and asked the Australian Ionising Radiation Advisory Council (AIRAC) to review radiation safety at test sites. The council appointed Dr G Watson, formerly of the Australian Atomic Energy Commission (AAEC), to undertake the review. Watson was an expert with considerable experience in radiation protection. But was the government falling back on the old stratagem of appointing an expert 'skilled in the area'? The AIRAC and Dr Watson had been close to scientists responsible for safety at test sites. The previous Liberal government had rejected veterans' claims largely on the advice of the AIRAC.

Watson's findings were as the veterans expected. Radiation

protection measures had been 'well planned and almost certainly effective'. There was no evidence of any serious unrecorded exposure to radiation.[24] Instead of quelling controversy, the government's choice of expert, as well as his findings, brought it to the boil.

When the government found the controversy would not simmer down it appointed a scientific review committee more removed from the past events. The committee found the AIRAC report presented 'a comfortable picture of the British nuclear tests' and that a royal commission was needed to probe radiation protection at test sites.[25]

The AIRAC publicly rejected the fairly clear implication that Watson, or the council, had favoured scientific colleagues in charge of test safety. The AIRAC responded by accusing the review committee of 'raising doubts about the competence and integrity of the Council and misrepresenting information in the report'.[26]

The debate was no longer seen to be just between veterans and the authorities; it was now being heard publicly from inside the scientific community. The government had no alternative but to appoint a royal commission. Royal commissions must work within the terms of reference set by governments. The deeper public concerns can be blurred and public debate closed off mid-stream. That is sometimes the hope of governments. However, with the more fruitful ones, legal logic has helped to clarify the issues. Cross-questioning has probed scientific objectivity to uncover, at times, professional ego and self-interest. This royal commission threw much light on the happenings on nuclear test sites. It did less for the veterans' cause.

THE Royal Commission into British Nuclear Tests in Australia was set up in July 1984, under the presidency of Jim McClelland, formerly a judge of the New South Wales environment court.

The commission found that the AIRAC 'had failed to make adequate inquiries before offering its conclusions'. This may have been because it was restricted by political pressures exerted by the government. The 'AIRAC with one exception spoke only to persons with an interest in advancing the view that the safety measures taken were adequate and effective. This led to an apparent bias in the material before it. As a consequence the AIRAC report cannot be taken as an objective and impartial assessment of the situation'.[27]

Contrary to the findings of the AIRAC review, the commission found that exposure to radiation at test sites had 'increased the risk of cancer among nuclear veterans'. Serious lapses had occurred in the enforcement of radiation protection at test sites which would have contributed to the risk. The absence of adequate records of radiation exposures, however, meant that the commission was unable to determine the degree of increased risk.[28]

Despite the findings of the commission the government moved slowly to compensate veterans suffering health damage. Three years later, McClelland had come around 'to the conclusion that the whole idea of the Royal Commission was to appear to be doing something'.

'I think it is government policy to sit and wait for 15 or 20 years until there's none of us left', Rick Johnstone reflected. 'We are down in numbers now and we're all getting older.'[29] Only when the British government moved to compensate its veterans for radiation-related injury did the Australian government announce compensation for a number of veterans.

The codes of radiation practice at test sites, according to the Royal Commission's findings, 'were reasonable and compatible with international recommendations at the time'.[30]

Patrick Green, on behalf of Greenpeace, had submitted to the commission that the ICRP recommendations, on which the codes were based, were compromised by conflicts of interest of ICRP members.[31] The nuclear establishment has always had close professional ties with the national bodies like the AIRAC and international bodies like the ICRP. Scientists who helped develop nuclear weapons and nuclear power sit on committees recommending radiation standards. The connections are uncomfortably close for disinterested decision-making on radiation safety.

Throughout the nuclear weapons testing, members of neither the AIRAC nor the ICRP joined with the concerned scientists who were publicly warning of the dangers of fallout. Yet by the end of the century the fallout from the tests could have caused, worldwide, over a half million people to die prematurely from cancer.

'In spite of its usefulness in the past,' Dr Karl Morgan (a founder of the science of health physics and once a member of the ICRP) said, 'the ICRP has never been willing to offend the establishment and I'm not sure it's an organisation that I would trust my life with'.[32] Yet that is just the situation in which the veterans were placed.

'There is a subtle source of bias that applies to all scientists who work every day around sources of radiation exposure, and that is that they have concluded that whatever risk they may be running from inevitable exposures, it is worth it to them to have their careers in this interesting and generally well-paid work', Edward Radford submitted to the Royal Commission. 'I maintain that, despite their technical knowledge of the subject, this potential bias makes them *less* appropriate as arbiters of what risks *others* should accept from radiation exposure, than scientists equally technically qualified but whose work and livelihood are not so directly related to the day-to-day use of radiation.'[33]

The Royal Commission in its conclusions did not concur outright with the submissions of Patrick Green and Edward Radford on the bias of ICRP members when setting radiation safety standards. The commission left it an open question, saying it could not decide whether or not the ICRP recommendations on radiation safety 'had been affected by the involvement of some of its ICRP members in the nuclear industry'.[34]

Chapter 9

The fallout legacy

> *Question*: How far away was the cloud when you first saw it?
>
> *Answer*: Kanytji said that when he saw the smoke it was not close, it was far away, but it was wide, it was fairly low on the ground and it looked black and spreading out.
>
> *Question*: Was the cloud moving?
>
> *Answer*: Kanytji said there was not much movement in it, that it was coming slow with the wind pushing it.
>
> *Question*: When you saw the cloud had you ever seen a cloud like that one before?
>
> *Answer*: Kanytji said that when he first saw the cloud, he did not see any cloud that looked like it before, and he said that they started to get a bit worried about it, they were frightened, because it looked black.
>
> — *Kanytji at the Royal Commission into British Nuclear Tests in Australia, answering through an interpreter about the black mist at Wallatinna, Transcript, p. 7183.*

IN the pre-dawn hours, ship's engineer Shinzo Suzuki stood at the stern of the tuna fishing boat *Lucky Dragon*. Unable to sleep, he had come on deck. Since sailing from Japan over a month earlier, he and his crew had had only poor hauls. Now they were trying their luck far from home, in the heart of the Marshall Islands.

Suddenly out of the murky blackness appeared a brilliant whitish-yellow flare in the distance. There followed a fireball rising above the western horizon, lighting up cloud cover and all around. Suzuki rushed into his cabin crying to a mate, 'The sun rises in the west!'. The sailors came on deck to see a flaming orange fireball rising over the horizon just like a sun. 'It can't be the sun in the west!', a seaman cried. Then the truth dawned; someone in a hushed voice said 'pica-don'. Minutes later the ship rocked as a great shock wave passed by, followed by great booms like cannon fire. Towering clouds billowed up from where the fireball had been. The Japanese seamen were uninvited observers of *Bravo*, the United State's first hydrogen bomb test on 1 March 1954.

The light of day calmed the sailors' fears. Then, as they hauled in lines for a hasty departure, a grey talc-like substance wafted down over the deck. Their skin became inflamed and their eyes smarted. By nightfall they felt nauseous. Seasoned seamen seemed strangely seasick on a calm sea.

Yomiuri's banner headline announced the homecoming to Japan of the tuna fishing boat. Doctors diagnosed the crew's ulcerated skin, falling hair and other maladies as symptoms of radiation sickness. The news reawakened Japanese fears of radioactivity. Health officials visited fish markets to find what they expected: Geiger counters near tuna from the *Lucky Dragon* clicked 30 000 times a minute instead of the normal 30. In the following months, other fish hauls were found to be radioactive. Three months after the *Lucky Dragon*'s fateful exposure the ship's radio operator died.[1]

In response to worldwide alarm, the USAEC explained how the radioactive level of the cloud and its fallout decreased very rapidly; most of the cloud's initial radioactivity had been lost by the time the fallout came down over the *Lucky Dragon*. That was true as far as it went. Most radionuclides formed in a nuclear fireball are extremely short-lived; many have half-lives of only minutes or hours. But that says nothing of the radionuclides remaining. On that the USAEC was silent (see Table 9.1).

The same brilliant light seen by the ship's crew flashed across Rongelap Island. The talc-like ash also wafted down over the island and its inhabitants. It fell all day, drifting into houses. The children played happily with their story-book snow. Not far across the sea from Rongelap, US technicians on Rongerik Island monitored the weather for the bomb tests. They retreated into a building tightly sealed from the fallout. Next day a naval ship took them on board, but did not rescue the islanders. Three men clad in protective clothing landed on the island and sampled the drinking water in the wells. The men left without saying anything to the native people. They and their children were left barely clad for two days amidst centimetres-thick radioactive 'snow'. By the next day the islanders were suffering acutely from their radiation exposure with nausea, vomiting, diarrhoea and smarting eyes.

Only on the third day were the islanders evacuated. 'When we arrived on Kwajalein,' Marshall Islander Etry Enos recalled, 'we started getting burns all over our bodies and people were feeling dizzy and weak. After two days something appeared under my fingernails and then my fingernails started coming off.'[2]

Until 1984 the Marshall Islands were a United Nations protectorate. Since 1946, Bikini, Enewetok and other Marshall Islands

Table 9.1: Important fission products in the fallout from nuclear explosions

Radionuclide	Half-life
Iodine-131	8 days
Barium-140	13 days
Ruthenium-103	39 days
Strontium-89	51 days
Yttrium-91	59 days
Zirconium-95	64 days
Cerium-144	284 days
Ruthenium-106	367 days
Caesium-134	2 years
Promethium-147	3 years
Strontium-90	29 years
Caesium-137	30 years
Samarium-151	90 years
Technetium-99	210 000 years
Zirconium-93	1 500 000 years
Caesium-135	3 000 000 years
Iodine-129	16 000 000 years

Source: J. Rotblat, *Nuclear Radiation in Warfare*, Stockholm International Peace Research Institute.

had been rocked by 66 nuclear explosions. Their inhabitants had been made refugees on other islands.

News of the *Lucky Dragon* awakened the world to the dangers of radioactive fallout. It rekindled a worldwide campaign to 'ban the bomb'. But for a long time the US 'protectors' allowed little or nothing of the truth to get out about the exposure of Rongelap Islanders to the fallout. In fact, the USAEC press releases said that the test of its first hydrogen bomb went routinely: evacuations of island inhabitants were carried out as a precautionary measure; there were no burns and all the people were reported to be well. It is true that, until the *Bravo* test in 1954, the island people downwind of the test sites had been evacuated before each test. But when *Bravo* was fired they were not evacuated; they were not even warned to take shelter. Yet the bomb was a thousand times more powerful than any previous bomb exploded. Its outer casing had been designed to yield extremely 'dirty' fallout.

All along, the test scientists have insisted that the fallout over Rongelap was an accident caused by a sudden shift of wind just as the bomb was fired. The weather technicians on Rongerik believe they know better. 'The wind had been blowing straight at us before the test', senior weather technician Gene Curbow said. 'It was blowing straight at us during the test and straight at us after it.'[3] The weather station had warned the test scientists before they fired their bomb that strong, high winds were headed for Rongelap. Curbow broke his silence after twenty years to tell the story. He suffers from cancer, which he attributes to the fallout.

The people of Rongelap were returned from Kwajalein to their island. Medical attention was lavished on them as on no other dependent people. At least once a year for 30 years, scientists from the Brookhaven laboratory came in a hospital ship to examine them. The ship brought all the scientists' food. They did not eat the choice fish, lobsters, clams, turtles and coconuts freely offered to them. In 1966, when scientists ate island food, their strontium-90 and caesium-137 levels had increased over a hundred times. After that the visitors politely refused all offers of the island's tempting delicacies.

Rongelap women experience a higher rate of stillbirths, miscarriages and birth deformities than other Marshallese women.

The women speak with anguish about 'jelly fish' babies. A woman described how she gave birth 'to something that did not look human — like the innards of a beast'. Over half the population of Rongelap suffer from hypothyroidism. One died of cancer, another of leukaemia; both can be linked to the fallout. Out of 22 Rongelap children exposed to the fallout, thyroid nodules were removed from 19 of them.

Lejok Anjain, a year old when *Bravo* was fired, died as a teenager of myelogenous leukaemia. He was taken to the United States, where he was medically examined in all kinds of ways. His mother believes they used him for their own scientific purposes. She said the people who brought the plague down on her son and the people 'think they are smart. They are clever at doing stupid things'.[4]

A Brookhaven laboratory report said, 'Greater knowledge of radiation effects on human beings is badly needed. The habitation of these people on the island will afford most valuable ecological radiation data on human beings'.[5] But the islanders became suspicious of the regular visits of the hospital ship. They gathered that the scientists were not really as interested in their well-being as they were to find out about the effects of radioactivity. They were being used like guinea pigs.

The people asked to be evacuated. The United States refused. In 1984 the Greenpeace ship, *Rainbow Warrior*, on one of its last voyages before being sunk by French secret agents, rescued them from the 'island laboratory'. They were taken to Mejato Island, where they are now rebuilding their lives. The United States has since tried to persuade the Rongelapans to return to their home island. This they refuse to do, until they know it is safe. That may not be for hundreds of years.

The downwinders

'It would be desirable if sites could be found which are so remote from populous areas that the tests could be conducted without regard to the direction of the winds. Unfortunately,' two nuclear scientists lamented, 'the bombs are too big and the planet too small.'[6]

Fallout from nuclear explosions may be 'dirty' or 'clean'. The nature of fallout depends on the bomb's composition, the power of the blast and its proximity to the ground when it explodes. Mushroom clouds rising out of nuclear fireballs carry aloft 'unburned' uranium or plutonium. When bombs are fired close to the ground huge quantities of earth are sucked into the fireball, where it becomes radioactive. The heaviest debris falls back in the vicinity of ground zero. Clouds of finer material will drift at low altitudes before depositing a hundred or more kilometres downwind. This is *early* fallout, which has about half the radioactivity.

The finest material rises into the stratosphere where it can drift around the globe for several years. Sporadically it returns to earth in rain. This is *delayed* fallout. Many tonnes of long-lived radioactive substances from nuclear weapons tests now impregnate the planet's ecosystems. Every person on the globe has in her or his body strontium-90 which, like plutonium-239, did not exist in nature before scientists split the atom.

Since the first nuclear explosion in 1945, over 400 have been fired in the atmosphere. The great majority were fired in the northern hemisphere by the two superpowers. Most of the fallout comes down in the hemisphere in which the bomb is exploded. Britain, France, the United States and very recently South Africa have fired a smaller number of bombs in the southern hemisphere. Since the 1963 partial test ban treaty most nuclear tests have been conducted underground. Between them the United States and the Soviet Union have exploded nearly one thousand nuclear devices in underground tunnels. Huge quantities of radioactive debris from these tests remain buried in the earth. Unknown quantities of radioactive gases have been vented to the atmosphere during the tests. The radioactive debris from French nuclear tests at Mururoa Atoll has escaped into ocean waters.

NATIONAL leaders on both sides of the cold war knew when they began testing nuclear weapons that fallout would damage the health of their own, as well as other, citizens. Albert Einstein had warned the world about 'the slow torture of radioactive dust and rain'.

Children can contract cancer and leukaemia from the strontium-90 and caesium-137 they absorb from the milk they drink. Children can be born malformed from carbon-14 in the body of their parents or even grandparents. Many statistical estimates with widely varying results have been made of how many children suffered as a result of the tests. But children and grandchildren suffering the consequences of the tests are not just statistics. Each was a beloved child. They are the unknown children sacrificed on the altar of national nuclear security. The nuclear establishment is forcing us to make a fatal choice — national security or a human future.

At the end of the 1950s public concern about fallout forced political leaders to declare a moratorium on open-air nuclear testing. The moratorium was observed by the two superpowers until 1961, when the Soviet Union abruptly started testing again.

French oceanographer Jacques Cousteau was an outspoken opponent of nuclear testing. He raised the matter of the renewed Soviet tests when he met a fellow oceanographer Dr Zennkevitch, president of the Soviet Academy of Sciences. 'Why all these bomb tests?' Cousteau asked his colleague. 'Did the Soviet government know what it was doing?' Zennkevitch responded with tears in his eyes. His academy had been consulted by the government about the fallout from the tests. Soviet scientists had told the government the tests 'would probably cause 50 000 children in the USSR to die. The answer was that if we did not test the bombs it may soon cost the Soviet Union many more lives'.[7]

AS the wind blows, so the radioactive cloud goes. Rain-outs can create 'hotspots' far afield. Chinese fallout has come down in North America, and Soviet fallout in Western Europe. In Australia, fallout has been recorded from British tests at Christmas Island, French tests at Mururoa Atoll and a joint Israeli-South African test in the South Atlantic.

At the nuclear explosions in the Monte Bello Islands, southerly winds were expected to blow the cloud in a northerly direction over Indonesia: better over your neighbour's backyard than your own. At Emu Fields and Maralinga, southwesterlies were chosen to blow the cloud over the land. The bomb-testers wanted to follow the clouds over the longest land route so they could sample them. The clouds were expected to pass out over the Pacific, north of Brisbane. Whatever the fallout, it was planned that it should come down over the sparsely populated outback and provincial coastal towns of north Queensland.

The reason for the chosen land route was explained by Professor Titterton of the weapons safety committee: 'It is sometimes necessary to make measurements considerable distances from the detonation centre; this is very difficult at sea.'[8] 'Sniffer' planes followed the clouds to take filter samples. Sticky papers and air filters were thinly dotted over the Australian landmass to collect radionuclides in the fallout. Radioactive iodine was measured in sheep thyroids.

These field tests were not for the purpose of monitoring the exposure of Australians to fallout; they were done to provide data on the behaviour of bombs during explosions. The data could also help British intelligence to identify pertinent information when they spied on Russian and American tests.

When weather and other natural elements are a factor in any human endeavour, the chance is that something will go wrong. This proved to be the rule rather than the exception for British nuclear testing in Australia.

By the time the Maralinga tests were to begin, fallout had become a sensitive political issue in Australia. Echoes of the *Lucky Dragon* incident in the Pacific had penetrated the federal cabinet. Mr Beale told the cabinet that since 'the Bikini hydrogen test in 1954, when several Japanese fishermen entered the prohibited area and suffered radiation injury, some public uneasiness has arisen in this country'.[9] The bomb-testers at Maralinga were instructed by the federal government to proceed cautiously and not to take chances on the weather as they had at Monte Bello. Whims of weather caused repeated delays of the first Maralinga test firing.

It was over two weeks before a favourable wind blew. Waiting impatiently, Lord Penney became frustrated and cabled the bomb-builders back home in Britain: 'Grateful for advice. Difficult for me to assess political strength of troublemakers raising scares by rainwater counts on Geiger counters. Can never guarantee that activity will not be found in rain 500 miles or more away'.[10]

During the British tests three-quarters of Australia, including populated areas, received fallout. 'Sniffer' planes tried to track the clouds, some of which went so far off course they were lost. Other clouds were traced thousands of kilometres off course. Instead of going over the sparsely populated areas, some clouds floated over the coast south of Brisbane. Fallout landed on the Victorian water catchment, and in Melbourne and Adelaide.[11]

The bomb-testers drew pear-shaped risk zones on their maps downwind of their test sites. Beyond the risk zones were safe zones, where likely radiation doses from any fallout were expected to be below the permissible limits. Any 'hotspots' during 'rain-outs' were accepted as unpredictable. Communities inside the zone boundaries were given no inkling of their place in the fallout plan.

THE stories of two communities tell us something about the miscalculation, deception and indifference to people's well-being on the part of scientists and governments bent on testing their nuclear devices, come what may. In appearance the two communities are worlds apart. One was the Mormon town of St George in Utah, in the south-west United States; the other was an Aboriginal community in the Australian red centre. However, both belong to arid regions sought out by bomb-testers.

St George lies 240 km away from ground zero on the Nevada nuclear-test range. Until the Chernobyl reactor accident in the Ukraine, the people of St George and Rongelap shared the misfortune of having their habitats made the most radioactive in the world in peace time. St George was dubbed 'Fallout City'.

The people of St George have always been strongly patriotic. They put only the word of their church above government authority. Officials coming from Washington played on this trust when they reassured the people about the safety of nuclear testing. The USAEC invited people to 'come and see history being made' at test sites. Thousands came to watch the nuclear fireworks. Teachers took their children to distant hilltops to witness the multicoloured mushroom cloud rising off the desert floor. Gamblers from Las Vegas casinos took excursions to watch the fiery dawn spectacles. Between 1951 and 1962 there were 87 of these open-air nuclear spectacles.

When the tests did not go to plan a grey ash fell over St George and other towns. It burnt people's skin. Scientists told the people that, if they stayed indoors, they would come to no harm. Officials were tight-lipped about the radioactive strontium and iodine in the fallout which contaminated milk supplies.

Irma Thomas lived in St George through all the bomb tests. Her husband is one of many who became afflicted with and died of cancer. Mrs Thomas's seven children are affected. She recalled how at the time 'doctors and scientists from Washington were coming here with assurances that we were not being harmed. They were urging us to build shelters because of the Russians. Well it turns out it was the Americans who were creating the danger. I shouldn't say this about our government but they lied'.[12]

Because of their abstemious living habits, Mormons generally have lower rates of malignancies than the national average. Today, the Mormon inhabitants of the town suffer from two to eleven times the number of cancers found among Mormons living remote from the test range.[13] The people in the towns and villages had had no need to look to medical records of their region to know that an epidemic of cancer had broken out among them. Families with long records of good health now found loved ones dying of leukaemia and cancer. In their small communities they were involved with each other's lives; and they suffered personally as they saw, for the first time, young neighbours die of leukaemia. It had all happened since the fallout, which must have been the cause of the tragic happenings.

The people were conservative and law-abiding. They became bitter and angry as the government fobbed off their questions with denials that the fallout had anything to do with the cancers and leukaemia. The community began organising support groups and fighting back. A number of individuals who suffered cancer, or whose relatives had been sufferers, sued the US government for radiation injury caused by fallout. The case was heard in a district court of Utah, presided over by judge Jenkins. In May 1984, he ruled that 'the defendant (the US government) failed to adequately warn the plaintiffs, or their predecessors, of known or foreseeable long-range biological consequences to adults and children to exposure to fallout from open-air atomic testing and that such failure was negligent. Instead of warnings,' the judge said, 'the USAEC's public information was heavy with reassurances'. He awarded some, though not all, claimants' compensation for their radiation-related injuries.[14]

In Jenkins's opinion, governments have a responsibility to take reasonable care to protect citizens' health and to inform them fully of potential dangers. The USAEC behaved negligently and deceitfully towards the downwind communities. The US Supreme Court overruled judge Jenkins's judgement on the grounds that, even though it may have been negligent, the USAEC had acted in the interests of national security. So justice for the downwinders was short-lived. The story of their battle is typical of the way in which social justice is usually denied to the growing numbers of innocent victims of technologies producing radioactive wastes, chemical pollution and food contamination.

American victims of toxic injury from radiation and chemicals resort more readily to litigation than do Australians, who show a preference for open judicial inquiry. The remedy in the long term lies in community participation in health and safety practices, and open and honest regulation of potentially harmful technologies. Compensation for toxic injury, resulting from biologically active agents released into the environment, and penalisation of those who cause it, should be a responsibility of society as a whole. Litigation cannot possibly cope adequately with the environmental and scientific complexities of toxic injury.

NO two court-rooms could be so different from each other as the one presided over by judge Jenkins in Utah and the one presided over by royal commissioner McClelland at Wallatinna in the red centre. But the issue was much the same: an uncaring government indifferent to the well-being of people downwind of a nuclear test site.

The royal commission hearings in the desert gathered evidence on the 'black mist'. Wallatinna was downwind when *Totem I* was exploded in 1953 at Emu Field 190 km to the south-west. At the time of the test 40 Aborigines were camped on the cattle station there. Wallatinna is now Aboriginal land. The hearings took place beneath an awning open to the scrubby red desert plain and stony hillocks. At the front sat the three commissioners. Assembled before the commissioners were ten or so barristers. Reporters sat at a press bench. In the midst of the court

Aboriginal witnesses sat in customary fashion cross-legged on the ground. They had been called in from across the desert by the Pitjantjatjara council radio transmitter in Alice Springs. In deference to the Aboriginal way of 'talking it out', justice McClelland ruled at the opening of the court that 'strict evidentiary relevance will not apply'.

Geoff Eames, counsel for the Aborigines, welcomed the Royal Commission onto Aboriginal land. 'Aboriginal evidence will be directed,' he said, 'specifically to establishing the existence of the black mist and to rebutting the AIRAC Report No. 9 which said that it was a proposition that lacked credibility'. The idea of a black mist had been 'treated with derision by Australian scientists'. Now a new analysis by British scientists, undertaken upon Aboriginal insistence, had demonstrated its existence. 'This Royal Commission is an inevitable result,' Eames said, 'of Aboriginal evidence of the black mist'.[15]

When the black mist *puyu* passed over Wallatinna, Yami Lester was only a child. At the time he was making his own fun filling an empty can with sand and dragging it along on the end of a wire. 'I was thinking it might be a dust storm,' Lester told the commissioners, 'but it was quiet, through the trees and above that again. It was just rolling and moving quietly'. The old people thought that it was a *mamu*, or spirit, which they tried to turn away with their woomeras.[16] Lester recalled how members of the community, including his own parents, Kanytji and Pingkayi, fell ill. His parents survived but others died. Lester's eyes became inflamed and, in later years, he went blind. These misfortunes began at the time of *puyu*.

When they tried to tell white officials about *puyu* their accounts were put down to fear and imagination. But *puyu* had been unique in the relationship the Aborigines had with all that had ever happened in their land; it was now a memorable episode in their oral history. Lester's mother Pingkayi remembered that 'the mist had got a strong smell like gas used in stoves. After the mist passed something like a frost was left behind on the leaves and ground'. Afterwards they had stomach pains; 'some people started coughing and vomiting and getting headaches. Pingkayi got sore eyes after *puyu* and now she has only one eye left and also Yami got sore eyes afterwards'.[17] Within days some people died and, as was the custom, the camp moved on. Soon they moved again after there was another death.

Yami Lester described to the commission how he became involved in the issue of the black mist. It was in 1980, the year he had been appointed director of the Institute for Aboriginal Development in Alice Springs. One day he heard Sir Ernest Titterton on the radio saying that Aboriginal safety had been taken care of during the tests. 'The scientist was talking a bit the wrong way', Lester said. 'Someone should answer him because nobody knew the Aboriginal side.'[18] Lester phoned a journalist friend in Adelaide to ask him to publicise his experience with the black mist. 'I don't know of any black mists,' Titterton retorted publicly. 'No black mists have ever been reported until this scare campaign started. An investigation would be a waste of money and time.'[19]

Eames asked Lester if he remembered how the AIRAC had 'dealt with the black mist and suggested that it was not a credible story'. 'Yes,' Lester replied, 'that was read to me and I did not agree with them for those clever people, they did not really want to help us'. He had proposed to the Pitjantjatjara Council that it call for a royal commission. Then they could tell their story, because 'in the circumstances we did not keep our records under the gum trees and the mulgas'.[20]

WHILE holding the inquiry in London, the Royal Commission had unearthed a report vindicating the Aborigines' story of the black mist.

For their calculations the AIRAC had relied on a model prepared by British scientists before *Totem I* was fired. The model was based on many assumptions: the power of the explosion, the mass of earth sucked into the fireball, the size of the particles in the cloud and how fast they fell back to earth, and ideas that certain shear winds would be blowing at different altitudes to disperse the cloud and its stem.[21] An underlying assumption seemed to be that God was on the side of the bomb-testers.

In 1983 two British weapons scientists re-evaluated the path the cloud from *Totem I* had actually taken. They had concluded that the cloud 'would have been a strange and awesome sight to anyone beneath it. A fine drizzle of black particles would also have been visible . . . to observers on the ground in the area of Wallatinna'.[22] The scientists had described *puyu*.

Having established the reality of the black mist, the Royal Commission sought to establish its health effects. Sir Edward Pochin, a radiation expert, was questioned on the subject. He had been involved with the ICRP from the 1950s and had advised on radiation practice at test sites. Geoff Eames questioned Pochin on the assumptions underlying current scientific understanding of radiation-related health effects. How was the actual amount of radiation absorbed by the human body estimated? What was known about how absorbed radioactive substances reached the gut? How much was needed to cause the nauseous condition described by Pingkayi? How much did people differ in their individual reactions to radiation?

Sir Edward Pochin somewhat reluctantly conceded that many untested assumptions are used when writing recommendations on radiation exposures. Little is yet known, he admitted, about how the state of health, lifestyle or genetic origins affect an individual's chance of suffering a radiation-related disease.[23] The differences could possibly be significant. These factors could have a considerable bearing on how radiation affected Aborigines. The expert had become defensive; certainty dissolved into uncertainty.

The commission concluded that 'the estimated dose to the gut from the black mist is close to the threshold for vomiting . . . Hence, it is possible that some individuals could have ingested enough fallout material soon after the passage of the cloud to produce vomiting'.[24] The commission thought a contributing factor to the Aborigines' illnesses could have been psychic. The Wallatinna community may have suffered 'a psychogenic reaction to a frightening experience', possibly in conjunction with the effect of radiation.[25] Scientists have speculated about stress exacerbating the biological effects of radiation. And the black mist was distressful to the Wallatinna community.

Whites reacted differently. Almerta Lander described the mist as it came over Never Never, where her husband was erecting a windmill. Its appearance was much as Yami Lester and his mother Pingkayi had described it. The mist had deposited an oily smear over her caravan and its furnishings. It left her with a strange impression but she had learned by word of mouth something 'was going on' down south.[26] The mist could be accommodated into white consciousness without too much unease.

But why single out the Aboriginal psyche? Why not the psyche of the bomb-testers at Emu Fields from whence the mist came? They had had unseeing eyes for a black mist visibly breaking away from their rising mushroom cloud. Radio reports, coming back from 'sniffer' planes, telling of a cloud drifting low over the desert floor towards Wallatinna, did not register with them. The scientists' denial of the black mist was not a cover-up. It simply did not fit the model — even though it was based more on assumption than hard data — and so it could not have happened. Were they under a spell cast by the fireball? One scientist who witnessed the first nuclear explosion at Alamogordo said, 'I haven't got over it yet. It was awful, ominous, personally threatening. I couldn't tell why'.

The portent of evil felt by the Wallatinna community came from their deep spiritual attachment to the land and its features. Many non-Aboriginal people have come to feel the same portent in radioactive clouds, less visible than the black mist, drifting out of control over their lands from nuclear catastrophes.

The Marston affair

Hedley Marston was chief of the division of biochemistry and general nutrition at the Commonwealth Scientific and Industrial Research Organisation (CSIRO). Marston was recognised internationally for his work on trace elements in soil. In 1956 he was asked by the Australian weapons test safety committee to monitor radioactive iodine-131 in the thyroid glands of cattle and sheep.

Marston first measured fine traces of iodine-131 that had wafted across the equator from nuclear explosions in the northern hemisphere. After the *Mosaic G1* firing at the Monte Bello Islands, in May 1956, he recorded a moderate increase in iodine-131 in slaughtered animal thyroids in the far north. *Mosaic G2* was fired in June. It was then that Marston became alarmed as he recorded a hundredfold increase on his earlier radiation counts. An 800-km swathe across northern Australia, he found, had been 'dressed' by a radioactive cloud. He reported his findings to the weapons safety committee.

Marston's findings were confidential. Not so the audible clicking of prospectors' Geiger counters. At Kuridala in far north Queensland, one prospector was startled to find the count from rainwater off his roof leap from the normal 15 to 2 000 counts per minute. A physics professor warned publicly against drinking water from areas of heavy fallout. Once again a Japanese ship found itself in radioactive fallout. This time it was a collier sailing off the Queensland coast. The cloud had drifted right across the Australian continent from faraway Monte Bello.[27] It was not the kind of publicity Canberra wanted.

Test scientists had advised the government that 'conditions for firing G2 were ideal from the point of view of safety of the mainland'.[28] 'What the bloody hell is going on? The cloud is drifting over the mainland.'[29] said a wire from acting prime minister, Sir Arthur Fadden. But that was not for public eyes. Marston was surprised when he heard Fadden reassuring Australians that the test 'was carried out without risk to life and property and absolutely without danger'.[30] In a private note to the head of CSIRO, Marston wrote saying his instruments must be faulty or else someone was lying. The irony would not escape anyone knowing Marston's scientific reputation and his forthright manner.

Other irritations came soon after. The committee reported that animal tests near Alice Springs showed no iodine-131. Marston knew otherwise. They must have done the tests at night when they could not read their instruments, was Marston's sardonic rejoinder. Then came a high count recorded in the south at his Adelaide laboratory. The count was close enough, Marston wrote, 'to contaminate the city and surrounding countryside'. The committee refuted Marston's claim, saying the cloud was never closer than 250 km. 'So much the worse for the inhabitants of the areas it did traverse', Marston said.[31]

He became more concerned about the long-lived strontium-90 than the iodine-131. 'The internal irradiation from these isotopes', he said, 'may after a latent period of years, result in many deaths from cancer.'[32] He was critical of the committee's reliance on radiation standards. They were, he said, 'founded essentially on guesses'. He knew from his soil work that much more had yet to be discovered about how strontium passed along food chains. His comments were rejected as exaggerated by the committee.

Ostensibly, the weapons safety committee had been appointed to act as Australian watchdog over the tests. But its members had set their minds on smoothing over any political obstacles in the way of the tests. They discussed how to present the bomb-testers publicly in the best possible light. Consequently, Marston's papers on his findings met with acrimonious criticism from the committee. The criticism was not devoid of self-interest. 'If a man of Marston's reputation was saying that these things [about radionuclides in fallout over Australia] were terrible, the tests were awful,' Lord Penney told the Royal Commission, 'of course there would be an almighty row.'[33] Marston had to be either silenced or discredited.

Marston had been attracted to the survey work out of a personal interest in soil trace elements. His disillusionment with the survey was undoubtedly due to seeing the cover-ups of his findings. But Marston may have also found himself at odds with his collaborators for philosophical as well as ethical reasons. Nuclear blasts are a brutal assault on the natural environment. Perhaps Marston came to feel his association with nuclear testing went against his life-long goal of determining trace elements to find ways to improve soil productivity. Dressing the land with poisonous radionuclides — as Marston with characteristic irony had labelled fallout — was a perversion of all that he had aspired to achieve.

Global fallout legacy

In 1957, while Marston was arousing the ire of his collaborators because of his expressions of concern about particular radionuclides in fallout, the American scientist Linus Pauling was facing impeachment by the United States Congress on the same score. He had published an estimate that tens of thousands of people would eventually die of leukaemia because of strontium-90 in fallout.[34]

Pauling also warned of the dangers of radioactive carbon-14 in fallout. He was joined in his warnings by Soviet scientist Andrei Sakharov. Carbon-14 exists naturally in trace amounts as the result of cosmic bombardment of atmospheric nitrogen, and is absorbed in the place of stable carbon-12 in living tissue. Pauling warned that the increase in carbon-14 in fallout 'will ultimately produce about one million seriously defective children and about two million embryonic and neonatal deaths and will cause many millions of people to suffer from minor hereditary defects'.[35] In 1962 Pauling was awarded the Nobel peace prize for his efforts to stop nuclear tests. He had already been awarded a Nobel prize for chemistry in 1954.

TRACES of tritium, a radioactive isotope of hydrogen, are widely distributed in nature. Tritium behaves like hydrogen and is also incorporated into living tissues. Before nuclear testing, the natural occurrence of tritium in rainfall had been extremely small. However, by the time most open-air nuclear tests had ceased in 1962, the count in the northern hemisphere had increased 2 000 times. In the southern hemisphere it increased 100 times by 1965, mostly as result of a slow drift across the equator.

In 1963 Ernest Sternglass published evidence linking infant mortality in the United States with radionuclides such as strontium-90 in fallout. Although the United States had ceased open-air tests, the USAEC feared that evidence linking infant mortality with radioactivity would damage the fledgling nuclear power industry. The USAEC sought to refute the Sternglass data; they asked their two most senior researchers in radiation biology, John Gofman and Arthur Tamplin, to review the Sternglass paper. The two researchers found that the estimates of infant mortality too high but that Sternglass's thesis was essentially valid.

Gofman and Tamplin were refused permission to publish the results without first putting a better face on nuclear safety. When they went ahead and published their findings, their supporting staff were withdrawn by their superiors. They resigned when it became obvious that there was no place in the USAEC for their open-minded approach to radiation and health.

WHEN scientists bored into the Arctic ice they found plutonium. They estimated that the first deposition occurred during 1955. Thereafter the concentration of plutonium increased up to 1963, when most open-air testing was halted. By 1979 almost all plutonium had come back to earth.

Nuclear testing has increased the natural radiation background by almost five per cent. Many radioisotopes from fallout will persist in the environment for thousands of years. Over four tonnes of plutonium have been deposited from fallout. It will persist for 240 000 years. Strontium-90 and caesium-137 will persist for over 300 years. These radionuclides are relatively insoluble but they are being slowly absorbed into food chains.

Fallout from weapons tests is only one of the many sources of exposure to radiation which result from *all* nuclear weapons activities. Production of nuclear weapons depends on the operation of the nuclear fuel cycle. Radionuclides migrate into the environment at every stage.

'The global victims of the radiation pollution from nuclear weapons production, testing, use and radioactive waste conservatively will number 13 million by the year 2000', according to Dr Rosalie Bertell. 'The current rate of weapons production in 1985 generates between between 7 000 and 15 000 victims yearly (between 20 and 40 a day).'[36]

'This we know', said chief Seattle, leader of the Suguamish tribe. 'All things are connected like the blood that unites one family. All things are connected. Whatever befalls the earth befalls the sons of the earth. Man did not weave the web of life; he is merely a strand in it. Whatever he does to the web, he does to himself.'[37] What we have done to ourselves from fallout has still to unfold.

Chapter 10

Radioactive wastes and wastelands

> A lot of people think, well, it's just going to be the Navajo that is going to be affected. So we shouldn't be saying anything, see. If something happens to us then its all right. But they don't realise that people eat the [contaminated] lamb chops in New York, Los Angeles, Miami and even London. Then they too have a part of that radioactivity.
>
> — *Kee Begay, a uranium miner suffering lung cancer, speaking in the film* The Four Corners: A National Sacrifice Area.

IN 1953, US President Eisenhower launched an international 'Atoms for Peace' program in the United Nations. He offered knowledge of the 'peaceful' atom in order that all might share in its abundant cheap and *clean* energy — an offer designed to divert world attention away from two problems of nuclear power: disposal of radioactive wastes and links between the peaceful and the warlike atoms.

Few scientists were prepared to speak publicly about the problem they saw looming of how and where to dispose of the radioactive wastes. One who voiced the scientists' dilemma was Wallace De Laguna, a health physicist working at the Oak Ridge laboratory, a major centre in the Manhattan Project. He argued in 1959 for the need to explain the problem in order that the public appreciate the 'peculiar precautions' needed to store the wastes safely. But he worried that people would see the wastes as a 'terrifying menace' and take 'unwise actions' to put a stop to the development of nuclear power.[1] In fact the first public action against waste disposal took place in 1959 when the Lower

Cape Cod committee on radioactive waste disposal campaigned against the dumping of the waste into the ocean 30 kilometres from Boston. In a letter to *The Bulletin of Atomic Scientists* members of the committee, writing in the capacity of 'citizens and scientists,' said 'strong opposition is expected from coastal areas'. With admirable understanding and foresight they went on to say that 'we believe the disposal of radioactive wastes will become the "number one" problem of the nuclear age'.[2]

Nuclear power was financed from burgeoning defence budgets. Scientists were enticed into the military fold through lavish funding from weapons programs. Already inclined towards the belief that technical problems were best solved by experts, the scientists allowed problems associated with radioactive wastes and their radiation emissions to remain hidden from public scrutiny. Had scientists spoken frankly about the radiation problem, people could have made up their minds much earlier about the merits of nuclear electricity. Instead, nuclear plants were promoted with expert assurances about safety, and were built in populous regions. Workers were told of 'safe' radiation levels.

Government nuclear regulatory bodies have played down the dangers of nuclear industry and have been reluctant to enforce stricter radiation safety standards. But there have also been scientists who have defected from the nuclear establishment to express their concern publicly about the inherent dangers of nuclear industry. Robert Pollard, of the US Nuclear Regulatory Commission, resigned saying, 'I could no longer, in conscience, participate in a process which so effectively evades . . . protection of public health and safety'.[3] The accounts that follow in this chapter illustrate the menace of radioactive wastes and the persistent and strenuous efforts made by citizens and scientists to bring the industry's problems into the light of day.

THE nuclear production system — uranium mining, enrichment, nuclear reactors, fuel fabrication, reprocessing and waste storage — is often referred to as the *nuclear-fuel cycle*. Facilities for these processes are scattered all over the world.

Australia's major link with the global nuclear-fuel cycle is supplying uranium to its 'front end'. Two uranium mines and two research reactors presently operate in Australia. A shipment of

Australian uranium may enter the nuclear-fuel cycle in the United States, Britain, the Soviet Union or Canada to be *enriched*. It may then pass on to a facility in one of a number of countries to be turned into *fuel rods* before being loaded into a *nuclear reactor* in a third country. After the uranium has been 'burned' in a reactor the 'ash', called *spent-fuel*, may be sent to either one of two *reprocessing* plants at Sellafield in Britain and Cap de la Hague in France. The plutonium is isolated during reprocessing. These two plants have been dubbed the world's 'radioactive rubbish bins'.

Uranium wastelands

The Colorado Plateau lies in the south-west of the United States. Its inhabitants have been affected by the new nuclear industry as much, or more, than any other people. On the plateau's western edge, nuclear weapons were tested in the 1950s in the open air. Now the tests take place underground. Downwind is the 'Fallout City' of St George. At Rocky Flats to the north-east, plutonium is prepared for nuclear weapons. Wind-blown plutonium dust from the plant contaminates the soil of the nearby city of Denver, where the incidence of radiation-related bone cancer is now rising. Almost half of all uranium produced in the world, up to the mid-1980s, was mined on the Colorado Plateau.

By a cruel twist of fate most of the world's uranium is being mined on the remnants of land left to the original inhabitants of the Australian and North American continents. Uranium mining is imposing on these indigenous communities chronic health problems caused by radioactive mine wastes.

To extract uranium the ore is pounded into a fine slurry in water. After extraction, a sand-like waste, called *tailings*, is run into a tailings dam. About three out of each 1000 kg of ore is uranium oxide. Tailings dams become hills of waste which retain over 80 per cent of the original radioactivity in the ore body. The radioactivity will persist for hundreds of thousands of years.

In the 1950s, uranium markets boomed with the new nuclear weapons industry. But uranium demand slumped once nuclear arsenals were filled. In the early 1970s, a burgeoning nuclear power industry sparked a second mining boom which was also short-lived. Jobs came and went with the rise and fall of the two booms. But the people of the plateau will be counting another cost besides lost jobs for a long time to come. A sacrifice of a land and its people lay behind a promise of wealth and jobs. The arid lands are now poisoned by radioactive dust spreading from eroding hills of mine waste. People on the plateau are suffering from an epidemic of cancer.

Most miners come from Navajo communities. Kee Begay mined uranium for many years. 'When I took the job they didn't tell me of the dangers', he said. 'Now look what it is doing to me.' Hundreds of Navajo like Kee Begay have contracted lung cancer.[4] Navajo miners are experiencing five times as much lung cancer as figures indicate they should. The primary cause of the miners' cancer is radioactive dust and radon gas. Radon atoms breathed into the lungs disintegrate to form other atoms which are also radioactive. They lodge in the lung tissues, where they emit alpha radiation at close range.

Cities, towns and villages on the plateau used the sand-like tailings to make concrete for street paving, schools and houses. Radon emanates from the contaminated concrete; the concrete is now being replaced, but this is a slow and costly business. Radioactive dust is blowing in from the desert. The incidence of lung cancer in the region is twice the national average.

Nuclear scientists should have warned people about the danger of building with the 'sand'. But how could they? People would have been awakened sooner to the dangers of tailings dust and of working in uranium mines. It was left to a growing number of cancers to flag the warning.

KEE Begay was not warned of the dangers by the radiation protection authorities who were supposed to protect him. Yami Lester, who suffered from the effects of fallout from a nuclear explosion in Australia, found the same reluctance among Australian scientists. 'Those clever people,' he said, 'they did not really want to help us.' Still today, warnings by radiation protection experts about radioactive dangers are usually muted.

Dr Joseph Wagoner, epidemiologist with the United States Public Health Service, has been more outspoken than most of his profession. When the mining boom commenced in the 1950s, Wagoner said, enough was known about the radiation hazards of uranium mining to insist on stringent safety measures. For over a century, 70 per cent of miners of uranium-bearing ores in Europe had died of cancer. Yet, up until the 1960s, regulations for ventilation of uranium mines were not enforced. That would have been true of uranium mining in the 1960s at Radium Hill in South Australia. However, even with improved mine ventilation today, uranium miners, according to Wagoner, still face 'twice the risk of developing lung cancer than do non-miners'.[5] That is the likely situation in the Roxby uranium mine in South Australia.

Meanwhile, in 1979, a tailings dam at Church Rock on the plateau collapsed. Millions of litres of water carried tailings down the Rio Puerto and out over its flood plains. When the Navajo people worried about their children playing in the river water, mining officials explained that, because the uranium had already been extracted at the mine, the spilled tailings were not a health danger. A deceit lay in what was not said. The danger from uranium mining comes mostly from the radium and other radionuclides remaining in the ore and tailings.

Navajo animals carry high levels of radionuclides which are swallowed with their food and water. The communities now hesitate to eat their own meat, because they are advised by health officials to avoid kidneys and liver from their slaughtered animals: it is these organs which accumulate most radioactivity. 'Even though you could say that the government regulations are not violated,' a health official said, 'we think you should look at ways to reduce your radiation exposure.'[6] In effect, no one should trust official 'safe' radiation limits.

ONCE open-air nuclear testing ceased, city dwellers forgot about radioactive dangers. Uranium mining occurs mostly in remote arid regions, and nuclear plants are mostly located away from large cities. But the biosphere is an interconnected whole. Waterways supplying Los Angeles and most of the south-west with drinking and irrigation waters pass through Navajo lands. The waters now carry three times the radium limit permitted by health standards. People in the cities have no option but to drink the water and eat the irrigated food.

Radionuclides are metabolised by plants and animals, thus

providing access to foodchains that lead to the meal table. Radium from mine wastes has been found to concentrate in aquatic creatures at levels up to 10 000 times higher than the levels in the water they inhabit.[7] Freshwater mussels in waterways running through uranium mining areas of the Northern Territory have been found to concentrate radium up to 20 000 times.[8]

Even before the world is far down the nuclear path — less than two per cent of the world's total energy supply comes from nuclear power reactors — we face the prospect of having to check our food for *radiation*, as well as *bacterial*, counts. People in the cities, said Kee Begay, 'will have a part of that radioactivity'.

IN the 1950s, Australia had its own uranium mini-boom to fuel British nuclear weapons. Mining took place at Radium Hill in South Australia and at Rum Jungle in the Northern Territory. A study of death certificates indicates that about one-third of Radium Hill miners died of cancer. Twenty-five years after mining at Rum Jungle, the Finnis River and its floodplains below the mine remain poisoned and infertile, despite costly efforts to rehabilitate the mine area. The riverlands have been abandoned by the Aborigines who once gathered plants, and fished and hunted there.

The Ranger mine in the Northern Territory began operating in 1981. Its tailings dam is poised upstream of the tropical wetlands of the Kakadu National Park. The fragile environment of the park's wetlands is no more secure than the wall of the Ranger tailings dam. Its collapse, like the collapse of the Church Rock dam, would mean inestimable damage to the downstream wetlands environment.

The Roxby Downs uranium deposit in South Australia is one of the world's largest. Its tailings dam could one day be the highest landform in the region. Plans to consolidate the tailings under a thin layer of soil appear a pitifully inadequate defence against the constancy of the desert winds and the longevity of the radionuclides. Over the centuries the sandy tailings will creep, like sand dunes, inexorably towards the wheatlands and the cities to the south.

A report by the Los Alamos scientific laboratory, a world mecca of nuclear research, sounds an ominous note. The environmental and health problems created by radioactive uranium mine wastes are probably irremediable. 'Perhaps the solution to the radon problem,' the report concluded, 'is to zone the land in uranium mining and milling districts to forbid human habitation.'[9] This is a grim prospect for Aboriginal communities and other Australians living near uranium mine wastes. And why should the wind-blown dusts stop at the borders of the scientists' zoned wastelands?

A radioactive rubbish bin

Each summer, people holiday along the coastal stretches of West Cumbria. Here the Cumbrian mountains and fells run down to the sea. Northwards across the waters of Solway Firth lies the Scottish coast, and to the west the Isle of Man rises out of the Irish Sea. Up in the mountains is the famous Lakes District. Though West Cumbria is not untouched by industrial activity, it still offers many havens attractive to holidaymakers.

Remoteness from populous parts has also appealed to the nuclear establishment for some of its dirtier activities. Sellafield on the Cumbrian coast is now the site of one of the world's large radioactive 'rubbish bins'. Another is located on Cap de la Hague jutting out into the English Channel from the coast of Normandy. A third is located at West Valley in New York State. This plant was launched by Governor Nelson Rockefeller as a symbol of imagination and foresight in nuclear enterprise. The plant became too 'hot' to handle and deteriorated into an embarrassing derelict for which no one wanted to take responsibility.

Other countries isolate plutonium from their own spent-fuel in small reprocessing plants. Since the breeder reactor program collapsed in the late 1980s, reprocessing to isolate plutonium no longer has any economic justification. Plutonium from reprocessing is an expensive fuel for power reactors, outpriced by uranium. Military use must now be the true purpose for a country to reprocess.

The remote, arid areas of the Northern Territory have been promoted as another dumping place for other countries' radioactive rubbish. Australia, it has been suggested, should take back the spent-fuel rods from its uranium customers' nuclear reactors; that, it is argued by some, would enable the plutonium to be safeguarded against use in nuclear weapons. Another effort to take Australia into the 'back-end' of the nuclear-fuel cycle is the development by ANSTO of Synroc, a high-level waste storage system.

THE Sellafield story offers some insights into the hazards of the highly radioactive wastes from the nuclear-fuel cycle. The first reprocessing plant was built at Sellafield (once known as Windscale) in the 1950s to produce plutonium for nuclear weapons. Later the plant was expanded to store and reprocess spent-fuel discharged from civil nuclear power stations. In the early 1970s Britain sought to gain a foothold in the promising world nuclear markets by offering to reprocess other countries' wastes and to return the isolated plutonium to customers.

Australia has long had links with Sellafield. Uranium from Rum Jungle and Radium Hill fuelled the first British power reactor at nearby Calder Hall. The plutonium in the spent-fuel from this reactor was isolated at Sellafield for British atomic bombs detonated at Maralinga. The spent-fuel derived from Australian uranium in other countries' nuclear reactors is now reprocessed at Sellafield.

The company controlling the Sellafield plant is British Nuclear Fuels Ltd (BNFL). The company is legally obliged to liaise with local bodies about public safety. Taking the provision seriously, West Cumbria Friends of the Earth tried to enter into consultations with plant management about its effluents. They were met with official disdain. Reprocessing was a matter for experts, not for meddling environmentalists. Consultation turned out to be nothing more than public-relations talks and guided tours in which the company told visitors about the harmlessness of its wastes pouring out into the Irish Sea.

For many years a million litres of radioactive effluent, containing plutonium, americium and forty other radionuclides, have been pouring daily into the Irish Sea two kilometres offshore. So far, 500 kg of plutonium have been deposited in the sea. Plant management seemed to expect the sea currents to disperse the wastes and carry them into the Atlantic.

The Irish Sea is now one of the most radioactive regions in the world. Sellafield's radionuclides are found in marine life along the Scandinavian coastline. Its plutonium is found in the Arctic ice. While some radioactive material migrated, some became fixed to seabed silt. Radioactive silt has been swept up onto Cumbrian and Scottish beaches and into river estuaries. Winds off the sea whip up the dried silt lying on beaches, carrying it onto farmlands and into houses many kilometres inland. Cancers in males in West Cumbria have risen from 10 per cent below the national average to 20 per cent above in the past decade. Autopsies show Cumbrians to have greater quantities of plutonium in their bones, sometimes thousands of times greater, than members of the general population.[10]

Radioactivity has spread to the south into the salt marshes of estuaries. Radiation-counts on estuary mud can reach a hundred times normal background levels. Sheep graze the area. Faraway city dwellers unknowingly eat the radionuclides with their meat. Samples of fish caught in the Irish Sea are contaminated with strontium-90 and other radionuclides. Irish fishermen are not allowed to sell their catch from the Irish Sea. The fish are weighed, paid for by an Irish government authority, and then burnt. Of course, the radionuclides stay in the ash or become airborne.

Radioactive ruthenium-106 concentrates in the seaweed used to make traditional Welsh lava bread. Since radiation-counts on the seafoods lay below the levels permitted by British food regulations, the contamination was not considered to be a health hazard. However, more recently the health authorities have shed some of their complacency as they have realised the significance, over a lifetime, for fish-eating people of the cumulative internal radiation-exposure of the ingested radionuclides. Cumbrians, like the Navajo, are now advised to limit their intake of locally produced food. In particular, they should not eat too many shellfish as they act as very effective filters of radionuclides. Many Cumbrians who loved to eat seafood now eat little of it or none at all.

The Merlins' house looks out over the Cumbrian coast. Dust in their vacuum cleaner has been shown to contain radioactive plutonium, americium, caesium and ruthenium. The origin of the dust is silt washed up onto the beach. 'There is this terrible menace,' Mrs Merlin said as she looked out from her porch across a stretch of the beach. 'It's insidious you see; you can't see it, touch it, smell it. You have no knowledge of where it is and what it's doing.'[11]

SEASCALE, a village of 2 000 people, is about three kilometres along the coast from Sellafield. The village and the Millom rural district in Cumbria have suffered an excess of childhood leukaemia and multiple myeloma at about eight times the national rate. More recently, an excess of child leukaemia of the same order has been found around another reprocessing plant which handles spent-fuel from fast-breeder reactors. The plant (now closed) is located at Dounreay in North Scotland.[12]

In 1984 a former president of the Royal College of Physicians, Sir Douglas Black, was appointed to investigate the high incidence of childhood leukaemia in Cumbria. Black found 'more cases of leukaemia in Seascale than there should be', and that the excess could be due to some local environmental factor. However, he concluded the estimated radiation dosage to the parish population was much too low to cause such a large excess of leukaemia. Black said, 'Common-sense dictates that there should be a connection but science is not common-sense'.[13]

But Cumbrians know from what scientists say that leukaemia is associated with radiation. The rise in leukaemia happened after the plant began its radioactive releases. The company has covered up accidents, and what it says about its releases cannot be trusted. Recent studies, such as those at Hiroshima, are showing radiation risks to be much greater than allowed for by Black. Some scientists say that we still have much to learn about the biology of radionuclides, such as plutonium, which emit the powerful alpha rays. Cumbrians have more plutonium in their bodies than anyone else. There is considerable uncertainty about the damage done by plutonium once in the human body. It seemed to make common-sense to link the wastes coming out of the BNFL plant with the excess cancer.

Excess cancers have been found in British communities situated downwind from other nuclear plants.[14] However, the picture was confused by a finding of excess childhood cancers in rural communities distant from, as well as downwind of, nuclear plants.[15] Since the Black report, a study of Seascale children shows the excess of leukaemia is occurring among children born in the locality and not born elsewhere.[16] Epidemiology, it seems, can see-saw for a long time before it settles down on one side. In the case of ionising radiation, the evidence is weighing ever more heavily on the side of its being more dangerous than previously thought.

In June 1988, the committee on medical aspects of radiation in the environment concluded that the Sellafield and Dounreay reprocessing plants are probably to blame for the leukaemias of children living nearby. 'You can never rule out chance but we have found an excess of leukaemia at the only two reprocessing plants in Britain,' said Professor Martin Bobrow, who headed the committee. 'It is hard to imagine that you could see an excess around Dounreay and an excess around Sellafield and it is just bad luck . . . You now have to prove that the excess of cancer is not due to radiation.'[17] It seems some scientists are prepared to mix science with common-sense. Perhaps too, some scientists feel more compassion than others when deciding who should be given the benefit of the doubt. Professor Bobrow heads paediatric research at Guy's medical school. Could he have thought that we should take no chances with children at risk of suffering leukaemia?

There was now a prima facie case against the plant to be answered. But the evidence remained largely circumstantial. Finding a credible answer to a scientific problem has always depended a lot on the inspired forethought that leads to the right question being asked. Most research up to the time of the review by Bobrow had looked for a connection between the excess of leukaemia and radioactive contamination of the environs around the plant. Irradiation of the sensitive foetus by radioactive contaminants seemed a most likely cause. But BFNL management could point to the low levels of radiation resulting from its effluents, at least according to its records. Even taking into account the recently increased radiation risk estimates, based on the revised Hiroshima data, the released radioactivity could only account for part of the excess disease.

Meanwhile, a research team led by epidemiologist Martin Gardner had asked themselves a different question: Could it be that radiation received by male workers inside the plant caused mutation of their sperm, to give a higher risk of the offspring developing cancer? In early 1990, after five years' work, the research team reported to the British medical research council that children of male nuclear workers who received a dose of 10 milliSv or more six months before conception were six to eight times more likely to develop leukaemia than other children. In Britain, it is known that more than 2 000 workers, over half of them in the nuclear industry, receive cumulative doses of more than 15 milliSv each year.[18]

The finding, like that of Alice Stewart's in 1957 — which showed low X-ray doses to the foetus increased the risk of childhood leukaemia — is a landmark in the understanding of the hazards of ionising radiations. It has far-reaching implications for the nuclear industry, raising doubts about the continued operation of the Sellafield plant. It also raises questions about present medical radiation practice. An early survey of radiologists had found an excess of genetic damage among their offspring.[19] Gardner's finding emphasises the need for careful shielding of a male's gonads during abdominal X-ray exams, and for a lapse of time to occur before conception of offspring to avoid the involvement of any mutated sperm cells. Of course, this new evidence of the induction of a predisposition to leukaemia in the offspring is not the only genetic damage that can be done to immediate and later generations.

The finding by Gardner and his team has been questioned, because it appears to conflict with the absence of excess leukaemia in the offspring of survivors of atomic explosions over Japan. But few if any children would have been conceived during the traumatic times immediately following the atomic blasts, and while any mutated sperm cells were viable.[20] Sperm cells involved in conception are formed in the testes, not long before conception, by a process of rapid cell-division. It is at the time of division that they are most vulnerable to mutation.

The dosimeter badges worn by radiation workers record only their external radiation dose. Besides external irradiation, mutation of the workers' sperm could result from internal radiation emitted by ingested radionuclides that concentrate in the genital organs or the semen. Plutonium, for instance, is known to concentrate in the gonads. In the case of Japanese bomb survivors the radiation dose was almost entirely external.

'I am in no doubt operations of this size could statistically lead to cancers', a BNFL spokesperson explained. 'But I think we should bear in mind the fact that these statistical cancers are, as they say, *hypothetical* cancers, and have to be judged against cancers that occur naturally anyway.'[21] Unfortunately, statistical cancers become tragically real to the unknown victims.

Many hypotheticals are involved also in the estimation of the doses of radiation that cause a certain number of cancers. With the recent findings on the incidence of cancer around nuclear plants and its likely causes, radiation protection experts have gone back to their computer models to reconsider some of their hypotheticals. A lot of assumptions are made about how radionuclides migrate after they are put out into the environment. A lot more is known about their physical behaviour than about their chemical and biological behaviour in the environment and the human body. This is especially so for the extremely toxic plutonium and americium unknown before the nuclear age. Their alpha rays are not readily measured inside the body, and there is great uncertainty about their behaviour in human bones.

Even the simple physical behaviour of radionuclides is often misunderstood. Estimates of inhaled radionuclides can be wrong by a wide margin: a very simple example is provided by the way we breathe in contaminated dust. When fine dust drifts near the body it is retained by clothing. Then, because our warm body behaves like a chimney, the dust is carried upwards in a warm air stream over our face. This 'chimney effect' caused workers in one nuclear plant to inhale up to a thousand times more plutonium dust than measured by the plant's sampling devices located on walls some distance away from the workers. Ivor Brown, a worker at the plant, became aware of the anomaly. In 1983 he pointed out the error to the management. His reward was the sack.[22]

No matter how much washing and vacuuming we do, dust clings to carpets, furnishings and clothes. When children romp and play they stir the dust, which rises up in their bodies' surface airstream to reach their mouths and noses. Any radionuclides adhering to the dust will enter the children's respiratory system. It could be happening in houses around nuclear plants.

JEAN McSorley was invited to Australia by Greenpeace to tell Australians of her experiences living in the radioactive environment created by the Sellafield reprocessing plant.[23] She reported that over 50 km of beaches in the region of Sellafield are now radioactive. People simply do not lie on the beach any more. Parents are even afraid to let their children play on the beaches. The radioactivity is over 20 000 times greater than the fallout from nuclear weapons testing in the 1950s.

'You can have first-rate science in Oxford and Cambridge,' McSorley said. 'But if you have third- or fourth-rate attitudes to health and safety, according to how they treat the local population, then these things will happen.

'John Dunster sits on the ICRP that oversees radiation standards worldwide. He was the man who actually started the Sellafield discharges into the sea and he deliberately increased them in 1956 as part of an experiment on the environment. Sellafield is the most cowboy nuclear operation in the whole world! So here is a man doing these things in the nuclear industry and at the same time sits on the national radiological protection board in Britain and the ICRP. It is so incestuous and compromising a situation that we would like to see it stopped.

'Unfortunately there is nothing we can do about the Sellafield area now. It will be contaminated for thousands and thousands of years. Plutonium has a half-life of 240 000 years.

'We want to tell Australians and others whom we visit not to go down the nuclear path. We say that, wherever we go. Don't take the waste, as your politicians are saying you should, because they will use it as an excuse and they'll go on creating it. And a country like Australia is the last place on earth that needs the nuclear industry and all its problems of radioactivity', she said.

Locking up health data

In 1964 the USAEC decided on an epidemiological study of workers' health at its military nuclear plants. Unlike the appalling working conditions at uranium mines, the AEC enforced seemingly strict radiation management at its plants. Workers' radiation doses, measured by their dosimeter badges, were routinely recorded and urine analyses conducted for workers in highly radioactive areas. Information on radiation exposure of about 300 000 nuclear workers had been recorded from all military nuclear plants, dating back to the 1940s. Such a data bank provided the basis for the largest study ever undertaken on the relation between low doses of ionising radiation absorbed over a long period of time and cancer. The Hiroshima survivors surveyed in the Life Span Study had received almost all their radiation doses in the instant of the explosion.

The USAEC's motives for undertaking the study turned out to be at least as much political as scientific. An internal memo, written at the time, says the study would 'probably not confirm or refute any important hypothesis but should permit a statement to the effect that a careful study of workers in the industry has disclosed no harmful effects of radiation'.[24]

In 1964, Dr Thomas Mancuso, a leading occupational epidemiologist, was appointed to undertake a health-and-mortality study. Mancuso had pioneered methods for detecting excess cancer among chemical workers. 'It is generally conceded', an AEC commissioner noted, 'that he is one of the best occupational epidemiologists in the world'.[25]

The study was to examine workers' health records at four sites where a range of nuclear activities had occurred for up to 20 years. Because cancer can take up to 40 years to develop, Mancuso planned the study to be ongoing and extend over the lifetime of the workers. The Hanford site in Washington State was chosen for the first study. It was here that nuclear reactors and a reprocessing plant had produced the plutonium for the Manhattan Project.

Mancuso's relations with USAEC staff were harmonious at the start. Relations began to sour once Mancuso refused to allow his reputation to be used to serve the public image of nuclear safety – which the organisation was intent on preserving. The turning point came when Dr Samuel Milham, a public health officer, found Hanford workers appeared to suffer from excess cancers. Milham had surveyed the death certificates of the workers. He had shown the results to USAEC officials before publishing and they had requested him not to publish his findings: publication would be against the interests of the nuclear industry.

The USAEC turned to Mancuso to publicly refute Milham's

results. Knowing the health officer to be competent, and given that his own studies were not yet advanced enough to come to any conclusion, Mancuso did not believe he could fairly criticise Milham's findings. He refused. That soured the relationship. He was then requested to publish the preliminary findings of his own study. Again he refused, since any conclusions at that stage could only be ambiguous. 'Any analysis which did not meet the number of years required to induce the occupational cancer would lead to false negative findings that would be misleading and could be misused.'[26]

Finding Mancuso unco-operative, the AEC cut short his funding for 'administrative reasons'. 'As I found out under the Freedom of Information Act', Mancuso said, 'they wanted to set me up to deny compensation claims of workers injured by radiation. I feel they tried to make me a patsy'.[27]

Mancuso looked across the Atlantic to enlist the help of Alice Stewart and her co-worker, statistician George Kneale. Stewart had recently published her pioneering health studies on X-rays, which had created difficulties for her in her relations with the British medical establishment. Kneale had pioneered statistical methods for dealing with the effects of radiation over the whole life-span. 'I have always taken the great precaution', Alice Stewart told the Royal Commission into British Nuclear Tests, 'to work with an experienced statistician, because my task is not to deal with sums, but to understand'.[28]

The research team confirmed Milham's findings of excess cancers. They found from their statistical analysis that six per cent of cancer fatalities among Hanford workers had been induced by radiation. Significantly, the excess was found to have occurred at radiation doses well below the levels permitted by the existing safety standards. It was concluded that ICRP estimates of the risk of cancer fatality from radiation were between ten and 30 times too low.

Alice Stewart describes their statistical approach to the data as seeing 'what the numbers had to tell them'. Their statistics treated the living as 'not yet dead' and who could yet contract cancer over their lifetimes. Their use of the so-called 'proportionate fatality' statistical method has met with a good deal of criticism. However, their findings on radiation risks are in the same direction as the recent findings of the Life Span Study. Combing the data, they found younger workers were more likely to be posted to higher radiation areas. Since cancers take up to forty or more years to develop, the final toll of cancer among those younger workers will not be fully known until they complete their lifespan. 'If you incur an induction of cancer at say 90, you will have little chance of expressing it', Alice Stewart said. 'But if you incur it at 20 you have a fairly good chance.'[29]

The USAEC tried every means to keep the findings secret. Mancuso's university was asked by the AEC to confiscate his research papers. In 1977 Mancuso, Stewart and Kneale published the results of their analysis in the journal *Health Physics*.[30] The same year, the US Department of Energy (DOE), the successor to AEC in weapons development, allocated funds for further study on their nuclear workers' health to Dr Sidney Marks, a former employee of the USAEC and later of DOE. Marks and his co-worker, Dr Ethel Gilbert, found an excess of multiple myeloma and cancer of the pancreas among Hanford employees, but concluded that this might be due to chemicals in the workplace rather than radiation.

The data was locked away from independent researchers. 'You cannot get through that brick wall', Alice Stewart told the Royal Commission in Australia in 1985. 'We have tried ourselves. Of course it is futile; several influential friends applying on our behalf met with a refusal.'[31] However, that was to underestimate the perseverance of community groups in the United States wanting to have the whole radiation story revealed. The Three Mile Island public health fund took court action, under the Freedom of Information Act, and persisted for three years before DOE agreed, in 1990, to release the data to Alice Stewart and her research team at Birmingham University.

Chapter 11

Nuclear reactors: the catastrophic consequences of failure

> On May Day Pripyat was a ghost town. Parks, gardens and sandpits were deserted. Offices, shops and schools were locked. There was no sign of life in any of the town's apartments. No radios or televisions played: no lights burned in the windows at night. The town had been hurriedly abandoned. Washing flapped unattended on clothes lines, children's toys lay discarded, apartments left in disarray. Windows and exterior doors had been taped up, the apartment blocks had been locked by officials . . .
>
> — *Henry Hamman & Stuart Parrott*
> May Day at Chernobyl, *p. 10, 1987.*

OF all the assurances given by the nuclear industry on the safety of its operations, none has been more discredited than the one concerning reactor safety. People were told that the possibility of a reactor catastrophe was 'less than the chance of a meteorite falling on your head'. Statisticians engaged by the industry produced estimates purporting to show that the probability of a catastrophe occurring was less than once in a million years.

In fact, scientists working on the design of nuclear reactors were well aware from the start of the possibility of accidents. Edward Teller, a leading scientist in the Manhattan Project, who headed a review of reactor safety in 1947, told a US congressional committee: 'The various committees dealing with reactor safety have come to the conclusion that none of the powerful reactors built, or suggested up to the present, are absolutely safe.' He urged, however, that fear of a major nuclear accident should not stand in the way of the rapid development of nuclear power.[1]

Today, the worst fears of a major accident have been realised in the ruination of Ukrainian and Byelorussian communities and their fertile lands by the reactor disaster at Chernobyl.

'AN accident has happened at the Chernobyl power station and one of the reactors has been damaged', came an impassive newsreader's voice over Moscow television. 'Measures are being taken to eliminate the consequences of the accident.' The accident began in the early morning hours of 26 April 1986. A power surge led to failure in the cooling system. As the fuel temperature soared the reactor exploded, blasting radioactive contents of the reactor core high into the sky. Radioactive clouds billowed out of the burning reactor for days. The early fallout descended upon Ukrainian and Byelorussian farmlands and villages downwind. People were moved off farms and out of homes for the sake of their lives. Yet nothing beyond the ash-blackened area around the fire-ravaged reactor had been visibly changed. The countryside, the garden and breadbasket of the Soviet Union, remained serene and lush in the early summer warmth.

Clouds of finer radioactive dust and gases carried beyond Soviet borders, descending on much of Europe. The very finest of the radioactive dust rose high into the upper atmosphere, from where it circled the earth until coming down in 'rain-outs' over far-away North America and Japan. The quantities of radioactive material that poured from the ruptured Chernobyl No. 4 reactor was greater than from an explosion of a nuclear weapon.

At the scene of the accident, firefighters deployed their fire-hoses in a hopeless battle against the blazing rupture in the reactor. One by one they collapsed in pain. Irradiation from the nuclear reactions still occurring in the exposed reactor core, and its debris, was hundreds of times more powerful than from the Hiroshima bomb. At Hiroshima the radiation consisted almost entirely of penetrating gamma rays. The firefighters at Chernobyl were irradiated not only by an invisible shower of gamma rays, but also by intense beta rays emitted by short-lived, fiercely radioactive fission products. The beta rays penetrated only a little below the skin, but resulted in vicious skin lesions and ulcerations.

One patient's fate was dramatic. 'He was already showing early signs of severe radiation exposure', according to a doctor attending at Moscow Hospital. 'His sufferings grew from one day to the next. In his mouth and on his face were large black herpes simplex blisters, often the first symptoms of exposure. His skin literally broke down before our eyes. First the sensitive folds around the groin and under the armpits became red and ulcerated. Slowly these ulcers spread across the body. In a matter of days he was covered with red, weeping skin burns. He was barely recognisable towards the end. We administered morphine, constantly increasing the dose, but that did little to ease his misery. The membranes that had lined his intestines had eroded and he suffered severe bloody diarrhoea. He died about 12 days after the explosion.'[2]

At an international conference on the Chernobyl accident organised by the International Atomic Energy Agency (IAEA), in August 1986, Soviet officials described the sequence of events leading up to the disaster and the measures adopted to bring the burning core under control.[3]

According to the Soviet report, thirty-one people — mostly firefighters — died of acute radiation sickness. Twenty-two had been exposed to more than 6 000 milliSv of gamma and beta radiation. They suffered nausea, vomiting and diarrhoea. Their skin ulcerated and disintegrated. They died within a month of the accident. Forty-four victims who had absorbed between 4 000 and 6 000 milliSv also suffered haemorrhaging and skin ulceration, which spread in waves over the body. They suffered from damage to bone marrow and intestines, and from infections brought on by the breakdown of their immune systems. Within seven weeks of the accident seven had died. Victims who absorbed between 1 000 and 4 000milliSv experienced mild haemorrhaging, minor skin ulceration and loss of immunity.

Officially, 238 people other than the firefighters suffered acute effects of radiation exposure. But in 1988 the Soviet radiological committee reported to a conference in Kiev that about 4 000 had received 2 000 milliSv and about 50 000 had received 500 milliSv or more.[4] People who received around 1 000 milliSv would suffer symptoms of radiation sickness and their life expectancy would be shortened. The 500 milliSv dose is greatly in excess of the cumulative dose for a whole lifetime permitted by international radiation safety standards. Since the accident 600 000 people have been registered as having been 'significantly exposed' to radiation. Those registered were living at the time of the accident in the exclusion zone and the more heavily contaminated areas beyond it. Even so, not all the workers, soldiers, medical personnel and others who took part in the clean-up and reconstruction efforts have been included on the register.

Many who fought the fire are now fighting for their life. 'It's been tough', said Yuri Hilko. 'Especially at first when I learnt I'd received 120 rems [1 200 milliSv] of radiation and later when my friends started dying.' Yuri Hilko is chronically ill and receives regular treatment for radiation sickness. Vladimir Usenko received 400 rems [4 000 milliSv]. People exposed to this radiation level will suffer debilitating illnesses and have their life shortened. Several times a year, when he feels seriously ill, Usenko goes to a radiation institute in Kiev for treatment. Once a year he goes to Moscow for blood transfusions and injections. 'It's thanks to the doctors I'm still alive', said Usenko, 'and I'm hopeful of the future.' His resigned manner belied his optimism.[4]

Immediately after the accident the authorities declared a 30 km exclusion zone around the crippled reactor. Four years later the zone was increased to 100 km. The towns nearest to the power station are Pripyat, where most power station workers once lived, and Chernobyl, the administrative centre for the region. In Pripyat, rows upon rows of apartment houses, schoolrooms and public buildings stand empty. Strenuous efforts to dislodge the fallout by hosing and scrubbing have failed to make the buildings safe. Chernobyl has been razed. The town's buildings would have remained too radioactive during their habitable life to be occupied.

Although the exclusion zone has lost its inhabitants, it is not deserted. A ceaseless effort must be made to stop the contamination spreading. The maximum permissible radiation dose was relaxed to cope with the emergency. Radiation levels near the reactor, immediately after the accident, were so high that workers had to be rotated every few minutes. After receiving 250 milliSv they were sent out of the exclusion zone for a year. The work of cleaning-up must go on for years, and the damage to the health of the workers will inevitably be great. They work a cycle of two weeks in the exclusion zone and two weeks 'holidaying' with their families. Much of the work has fallen to soldiers who have been detailed for turns of duty in the area to hose down roads, trees and buildings, and demolish contaminated buildings.

The three remaining power units were decontaminated. Wiping down with solvents, scrubbing with water, and spraying concrete with plastic were among the methods used to lower the radiation levels to which power workers would be exposed. 'Decontamination' has become the magic word officials like to use to assure people that the radiation danger has somehow evaporated. But at Maralinga decontamination of vehicles failed dismally. At the Three Mile Island power station in Pennsylvania, where a reactor core partially melted in 1979, several years of costly decontamination has left concrete still exuding radioactive materials.[5] In 1990, the Ukrainian government decided to have the Chernobyl power station closed entirely by 1995.[6]

THE Ukrainian city of Kiev lies only 80 kilometres to the south of Chernobyl. For several days after the accident it was business as usual in Kiev. The May Day parade was held with the usual colour and pomp. Then the wind changed and fallout descended. Calm gave way to panic. People fought to send their children away on trains, buses and planes beyond the reach of the fallout. Days before, local party leaders in the know had sent their children to safety. Playgrounds and schools emptied. Kiev became a city without children.

Doctors fearing genetic deformities advised pregnant women to have abortions. Even when the dose during pregnancy is too low to cause outward deformity, radiation can cause mental retardation in the child. Health authorities exhorted people to battle against 'our enemy the dust'. Several times a day pavements were hosed and buildings doused with water. Washing clothes became a preoccupation. Wet rags were draped at building entrances for people entering to wipe away the dust clinging to their shoes.

Newspapers devoted much space to what people might 'safely' eat. They described in great detail checks made on fish from the Dnieper River. They would report not only whether the fish were contaminated but which parts — the gills, internal organs, fins or tails — showed the least traces of radioactivity. The authorities insisted farm produce be checked with a Geiger counter. 'All checked, all checked', stall-holders would call in food markets. If the count went too high the food was rejected. But the shoppers were not so easily reassured about radioactivity. The authorities had failed earlier to warn of fallout dangers; they were no longer to be trusted. People washed the vegetables they bought with a hose. Of course, that only put the radioactivity somewhere else.

All efforts of the Soviet authorities to create a feeling of normality are now mocked by a huge sarcophagus of concrete and steel looming over the exclusion zone. Nuclear reactions inside the sarcophagus cannot be entirely repressed; it has been designed to be constantly cooled by an air-draught passing through its concrete shell. It must be tended for hundreds or possibly thousands of years. No one knows.

For people in the farming areas of Byelorussia and the Ukraine, the consequences grow worse year by year. In July 1989 Soviet authorities decreed that, in areas where a *lifetime* radiation dose would exceed 350 milliSv, inhabitants would be resettled over the following three years. After visiting the contaminated areas, a Red Cross team decided that, on the basis of this maximum lifetime dose, at least another 100 000 people would have to be resettled.[7] The maximum dose recommended by the ICRP is one milliSv per annum, or about 70 milliSv over an average lifetime. Thus many people will still be left exposed to radiation normally considered unacceptably high. Eventually, a quarter of a million people or more will be resettled. In March 1990 parliamentarians from Byelorussia and the Ukraine appealed to the United Nations for help to solve the massive health and food supply problems still confronting their people because of radioactive contamination. Over two million people are living in contaminated areas. They estimated that the cost of cleaning up completely amounts to 200 billion roubles ($US100 billion).[8]

And the economic cost will be the least cost. In the Soviet Union and Europe, a shortened life because of cancer is the inevitable fate for many unknown victims. Estimates of fatal cancers resulting from the fallout range from official estimates of around 10 000 up to 500 000; and as the years go by the higher figure appears the more likely.[9] According to then-West German geneticist, Helmut Hirsch, in his country alone 'it is expected that there will be from 4 000 to 23 000 additional cases of cancer (apart from thyroid cancer) and 90 000 hereditary disorders'.[10]

A few days after Europeans learned of the accident, Dr Auril Arthur went walking with a Geiger counter in a Munich park. As she watched children playing on the grass she wrote to a newspaper: 'Had our lab, where we use radioactivity for research, been as heavily contaminated as the city of Munich this sunny May holiday (and much of Eastern Europe is worse) it would have been closed immediately.'[11]

Across Europe people were warned not to consume contaminated vegetables, milk and water. People sought frantically to buy the remaining 'pre-Chernobyl' food. Huge quantities of contaminated food were buried. European Atomic Energy Agency officials argued that their own nation's contaminated food should be exempted from the trade embargo. Parents were told to keep their children indoors at home. But fallout from a nuclear accident is ubiquitous. Families were given stable iodine tablets to counteract radioactive iodine in their thyroids. In Poland, children drank radioactive milk because there was no other. Far away across Siberia, in Japan, radioactive iodine was found in a mother's milk. Sheep in Britain became too radioactive to market. Italian spaghetti was found to have excessive caesium-137 in it.

Radiation protection officials tried to deflect attention from their own nuclear industries; they announced that low-level radiation from the fallout was nothing to worry about. Ordinary people distrusted these officials, so they turned to Friends of the Earth, Greenpeace and other community environmental bodies for advice on how to protect children from contamination.

A LONG-LASTING legacy of Chernobyl fallout is radioactively contaminated food. Irish, Scottish and Cumbrian meat, Polish potatoes, Greek wheat, and Turkish tea became unmarketable overnight. Milk became undrinkable while still palatable. Italian pastas were found to have fallout radionuclides in them. The door closed on much of Eastern Europe's food exports. International competitors moved quickly into Europe's foreign markets, offering uncontaminated food. Australian wine sales boomed in Europe, because it was fermented from antipodean grapes.

In the post-Chernobyl era, countries insisted that radiation counts be taken on food imports from Europe — just like the dieldrin tests which are now performed on imported Australian meat. A device to dispose of stocks of contaminated food, which was uncovered by community activists, was to mix it with 'clean' food. This simply spreads the risk over a greater number of people while leaving the number of cancer victims unchanged. Greek officials sought to dilute the radioactivity of 600 000 tonnes of durum wheat in this way before turning it into pasta. Australian customs authorities have condoned this practice with some imported food.

Radioactive food has turned up and been impounded in the distant ports of Alexandria, Penang, Colombo and Bangkok. Shipments of contaminated powdered milk, donated (or dumped) by the European Economic Community, have been returned by recipient Third World countries. The *Reefer Rio* sailed between Atlantic ports for over a year unsuccessfully trying to find a trader willing to pass off its cargo of radioactive Irish and Danish beef. It was returned to cold store in the German port of Wilhelmshaven to await unsuspecting customers. Chernobyl has given us a little taste of food in a nuclear society.

'Hotspots', many times higher in radioactivity than pre-Chernobyl, are being detected in Britain and Europe. Farm products from the western upland areas of Britain have been made unmarketable for some years. Reindeer meat in Scandinavia remains highly radioactive. Reindeer feed on lichens that selectively filter off radionuclides from the environment.

Fallout radionuclides in the environment do not stand still. Fish caught in Swedish inland lakes register high radioactivity. The radionuclides are migrating with natural drainage towards lower-lying lakes and marshes. As the years go by, low-lying areas will become more, not less, radioactive. The organic matter in vegetable and other crops that were buried will have decomposed, but many of the radionuclides contaminating the crops will live on for hundreds of years. The Turkish government bought contaminated tea from its farmers but then had trouble disposing of it. Burial in disused mines was resisted by local people, who rightly feared contamination of their ground water. Turkish officials would like to dump it into the Black Sea, but this is opposed by fishing communities.

RADIATION counting is much easier than testing for insecticides chemically. A well-equipped laboratory is a help but not essential. Spurred on by mistrust of official regulation, community groups have acquired radiation counters, sometimes assisted by home computers, to do their own monitoring. Relatively inexpensive counters measure not only radiation levels, but also reveal the actual radionuclides present in food. We can now know whether we are eating radioactive caesium, strontium, iodine or whatever.

The European Commission is wrestling with the prospect of food distribution where 40 per cent of the food supply will be contaminated; it wants to convince Europeans to tolerate higher levels of radionuclides in their diet 'in case of abnormal levels of radioactivity or of a nuclear accident'. The answer is not unexpected: raise the limits for an emergency period after an accident. The 'tolerance' limits of radioactivity in food for the next 'post-Chernobyl' emergency are to be at least twice as high as the upper levels normally applying (see Table 11.1).

The last stand of the nuclear industry in its battle for survival will probably be fought on the food front. Certainly, the nuclear industry is losing friends over what it is doing to their food. The proposed higher 'tolerance' limits face determined opposition from the majority in the European Parliament, which has voted to provide special categories of contamination limits for pregnant mothers, children and the sick. Uncontaminated food must be made available to these vulnerable groups. The Parliament, from its considerations on radioactivity in food, concluded that 'a responsible health policy can only be pursued if the use of nuclear energy is abandoned'.[12] If the European Parliament — which has included a number of 'green' representatives — has its way, food will be labelled with the levels of radionuclides present.

Then the staunchest advocates of nuclear power will find their loyalty severely tested in their own kitchens.

Table 11.1: The limits of radioactivity set by the European Commission for the common foods — and higher limits set in case of a nuclear accident

Radionuclides	Radioactivity Becquerels/kg	
	Standard	Emergency
	Milk	
Iodine-131	—	500 (130)
Caesium-134/Caesium-137	370	1 000 (100)
Strontium-90	—	(25)
Plutonium-239/Americium-241	—	20 (2)
	Other foods	
Iodine-131	—	2 000 (1 300)
Caesium-134/Caesium-137	600	1 250 (125)
Strontium-90	750	(150)
Plutonium-239/Americium-241	—	80 (8)

Note: The figures in brackets are those adopted by the European Parliament.

Our own little reactors

The Australian Atomic Energy Commission (AAEC) was set up in 1953, the same year that President Eisenhower launched his Atoms for Peace program. A primary concern of the new organisation was to develop industrial uses of atomic energy in Australia. Another concern, not publicised, was to gain the nuclear know-how needed should the government later decide that Australia 'produce plutonium for military purposes'.[13] Official records show that the weapons option remained a factor in the AAEC's investigations of nuclear reactors.[14]

The AAEC decided its reactors should be located 'remote from residential areas'. A site was chosen at Lucas Heights, an area of bushland and river valleys beyond the sprawling suburbia south of Sydney, though only 30 km from the heart of the city. Here the AAEC felt its nuclear reactors would bother no one. It built two research reactors, HIFAR and MOATA. In 1959 the larger heavy-water reactor HIFAR went critical; that is, the splitting atoms emitted neutrons at a sufficient level to sustain a chain reaction. MOATA, a light-water reactor, went critical in 1961.

In the early days of the Lucas Heights research establishment, as the laboratories later became known, few people had settled in the nearby bushlands. By the 1980s housing subdivisions had overtaken the greenbelt around the establishment. The AAEC decided, perhaps in an attempt to show how safe its activities were, that it no longer needed to keep its distance from residential areas. Housing development was encouraged up to the edge of a shrunken 1 600-metre exclusion zone. At first, residents in the district took no exception to their apparently inoffensive neighbour. Why should they? There were no offensive odours, no belching chimneys and no noise; and the AAEC saw no reason to publicise its unsensed radioactive effluents or the unlikely possibility of an accident.

Then, in 1979, a serious nuclear accident occurred at Three Mile Island in the United States. The world watched anxiously for five days while engineers coaxed the stricken reactor back from the brink of catastrophe. Had the very worst happened at Three Mile Island, a large area of the state of Pennsylvania would have been made uninhabitable by fallout. The suspense of those eventful days, when women and children were seen on TV screens fleeing from the damaged reactor, changed local residents' attitudes about their own nuclear neighbour. The even more climactic events surrounding the Chernobyl reactor explosion in 1986 further heightened residents' concern.

Of course, HIFAR and MOATA together have only one hundredth the potential for damage of the Three Mile Island or Chernobyl reactors. Even so, there are parallels to be drawn. The town of Chernobyl was 19 kilometres from the reactor, and Kiev 80 kilometres. Chernobyl was made so highly radioactive it had to be demolished entirely. Kiev's citizens were thrown into panic by the fallout, and they have since felt anxiety about genetic disorders and latent cancers.

The national health and safety committee assessed a catastrophe at Lucas Heights to be of low probability but of huge proportions. The HIFAR reactor is only '1 600 metres from the nearest resident', the resident action group pointed out, 'and within 80 kilometres of three to four million people'.[15] Chernobyl has shown that radioactive decontamination of residential areas, even when it is undertaken a long way from an accident, can be an arduous task. And anxiety about health can be long-lasting and well-founded.

Only after the Three Mile Island accident did the AAEC reveal that it had had an emergency plan since the early 1960s. The plan, code-named *Aptcare*, provides for evacuation of downwind residents. An abridged version of Aptcare, released publicly, says nothing about how people should respond to an accident; it talks mostly about why an accident cannot happen.

Residents asked health authorities to issue stable iodine tablets to counteract radioactive iodine released during an accident. The request was refused. Of course, the tablets would be a constant reminder that a serious reactor accident might occur one day. The reluctance of the AAEC to involve local residents reflects the dilemma of nuclear authorities. How can they cultivate public confidence while organising evacuation drills?

Insurance companies are apparently unconvinced that there will be no costly accident at Lucas Heights. They will not insure against damage by a reactor accident. The government hedges on compensation; it will only pay compensation awarded under common law 'in the unlikely event of such action being taken and being successful'.[16]

The Sutherland shire council agreed to the building of the AAEC reactors in its municipality on the basis of false assurances. In 1955, AAEC chairperson Sir Phillip Baxter told a meeting of the council that there would be no escape of radioactive material of any kind from the reactors. 'There is no gas, no liquid, as the whole is completely enclosed . . . No radioactive material would find its way into the river.'[17]

Research reactors, like power reactors, regularly discharge radioactive wastes — not nearly as much, but some. In a typical year the gases and vapours discharged from the stack at Lucas Heights have an activity equivalent to over 10kg of radium. Most radioactivity is associated with argon and other noble gases from the HIFAR reactor. This gas is chemically inert but radiologically active if breathed into the lungs. A wind study has shown that 'pollution from Lucas Heights may well affect residential areas nearby'.[18] The gases from the stack are supposed to rise into the upper atmosphere. However, with the frequent temperature inversions in the river valleys, gases would often hover over residential areas.

Strontium-90 was distributed over the earth's surface by nuclear weapons testing. Early surveys show that strontium-90 was higher in milk produced around Lucas Heights than further afield. The AAEC attributed the excess to uneven distribution of weapons fallout. A contrary opinion is that the excess strontium-90 comes from HIFAR. The same pattern of high levels of strontium-90 has been found around a number of power reactors overseas. The AAEC ceased its strontium-90 surveys in 1970.[19]

OVER the past 30 years radioactive wastes have accumulated in large quantities at Lucas Heights. The scant accounts of waste management found in the AAEC's glossy annual reports bear little resemblance to the contents of its internal environmental reports — in which, by contrast, we find accounts of a radioactive graveyard leaking radionuclides into a local creek, storm-water carrying tritium into oyster farms, and gases escaping over residential areas. 'The history of the operation of the establishment at Lucas Heights', the resident action group said, 'is characterised by misinformation, concealment and secrecy'.[20]

Low-level radioactive waste was buried at the Little Forest burial ground, now on the edge of the buffer zone. The burial ground was closed in 1970 when radionuclides were found to have migrated into nearby grasses and trees. The ground was fenced and top-dressed with soil. But radioisotopes have no respect for fences and they travel through soil. Errant radionuclides are still migrating into the sand and water of local waterways.

MOATA is cooled by ordinary water (light water). A pipeline poured millions of litres of mildly radioactive water into the Woronora River each month. Water weeds, fish and oysters have been contaminated with cobalt-60 and strontium-90. Because blackfish feed on weeds they accumulate more radionuclides than other fish. Since 1981 the water wastes have been piped out to sea as a result of residents' protests and petitions. As a signatory to the South Pacific nuclear-free-zone treaty, Australia is committed not to put radioactive waste, however diluted, into Pacific waters.

Highly enriched weapons-grade uranium is used to fuel the reactors. The spent-fuel rods withdrawn from the reactor contain weapons-grade plutonium. Over a thousand of these rods are now held in stainless-steel containers at the site. The residents action group has submitted to the federal government 'that a national repository should be established to store the large amount of low-, medium- and high-level waste stored on the site. The waste should be stored in a manner which presents the least risk to workers, the public and environment for all time'.[21] A national repository has also been recommended by the AIRAC. But where? Much smaller quantities of waste accumulated by other radioactive ventures have yet to find a final resting place. Who in Australia wants radioactive wastes in their own backyard?

Until 1987 the AAEC acted as its own referee on radioactive waste-management practices, when it was replaced by two nuclear bodies: the Australian Nuclear Science and Technology Organisation (ANSTO), responsible for nuclear research, and the Nuclear Safety Bureau (NSB), responsible for monitoring and reviewing the safety of ANSTO's nuclear reactor operations.

While the two bodies are administratively separate, they both appear likely to perpetuate the same technocratic ethos of the disbanded AAEC. In its 1968 annual report, the AAEC explained 'the need for balance between need for development of nuclear industry and its duty to protect population'. So far the balance is being struck by its successor ANSTO in a way that is being interpreted by residents as 'business as usual'. ANSTO clearly intends to continue to operate nuclear reactors alongside Australia's most populous suburbs.

Today the operation of ANSTO's nuclear reactors is justified largely by the production of radioisotopes (radionuclides) for nuclear medicine. However, a *cyclotron* can be used to provide isotopes, with medical advantages over reactor-produced isotopes and at much less cost. A cyclotron uses high-energy beams of atomic particles to convert stable atoms of a target into radioisotopes. Because the radioisotopes are short-lived, cyclotrons are best located at the hospital where the diagnoses are conducted. According to Dr John Morris, who heads a nuclear medicine department, cyclotron-produced isotopes are 'biologically superior to any made in a reactor because they are the isotopes of the building blocks of nature: oxygen, nitrogen and carbon'.[22] Because these isotopes are short-lived they expose patients to the least possible residual radiation after their diagnosis is completed. Reactor-produced iodine-131 gives a much higher radiation dose to patients than the shorter-lived cyclotron-produced iodine-123.

The Lucas Heights research establishment was set up in the 1950s during the heady days of the new global nuclear enterprise. Nuclear power has long since faded as an acceptable energy option for Australia. Medical isotope production no longer provides credible grounds for operating a reactor. A government review concluded that producing medical isotopes at Lucas Heights actually 'inhibited the most cost-effective isotope program'.[23]

Why then should the government persist with the heavy drain that the reactors make on the national purse? Nuclear energy swallows over half the energy research budget, while solar energy research lacks adequate funding. Could it be that the reactors are seen by defence policymakers as keeping open the nuclear weapons option? If so, then once again national security — or a false sense of its meaning — is putting at risk the health and safety of Australians.

isotopes. Because the radio-isotopes are short-lived, cyclotrons [illegible] be located at the hospital where the diagnoses are conducted. According to Dr John Mercer, the head of a [illegible] department, cyclotron-produced isotopes are biologically superior to [illegible] reactor [illegible], they are the isotopes of life itself—carbon, nitrogen and oxygen and [illegible] because they are short-lived they expose patients to less [illegible]. The [illegible] after their diagnosis is complete. Reactor-produced iodine-131 delivers much higher [illegible] dose to patients than the short-lived iodine produced in [illegible]

The Lucas Heights research establishment [illegible] set up in the [illegible] the [illegible] The new global nuclear [illegible] [illegible] since [illegible] an [illegible] energy [illegible] Australia. [illegible] [illegible] [illegible] for [illegible] A [illegible] review concluded that producing medical isotopes [illegible] only marginally influenced [illegible] reactor [illegible]

Why then should the government persist with the Lucas Heights [illegible] the reactor and [illegible] [illegible] [illegible] [illegible] [illegible] Could it be that [illegible] [illegible] policy [illegible] [illegible] [illegible] [illegible] [illegible] [illegible] national [illegible] [illegible] [illegible] [illegible] [illegible] McCallum.

PART III

The Health Research Issues

Radiation exposure from nuclear and non-nuclear sources is part of a wider public-health problem created by the release into the environment by modern industry of innumerable *biologically active* agents. The study of environmentally induced disease presents some very difficult problems: in particular, identifying the causative agent and elucidating the biological pathways leading to the disease.

Epidemiological studies of exposed populations so far provide the most useful measure of diseases caused by radiation. However, experts differ greatly in their interpretations of these studies and their meaning for public health.

Biological studies on cultivated cells are directed towards understanding how radiations damage human health. Progress is slow, and little is yet known of the pathways by which radiation-related diseases develop. In the case of health effects from non-ionising radiations, about which there is little common ground in the scientific community, some of the most fruitful laboratory studies have been on hormone-secretion rates and related behavioural changes.

While experts differ, communities are increasingly finding the need to become informed about the scientific basis of these health effects. They are no longer prepared to automatically accept radiation safety standards and the 'permissible' exposures prescribed in them. The essential information can be ferreted out, and communities are drawing their own conclusions.

Chapter 12

Biological studies

> ... The overall body of scientific evidence ... is in favour of the concept that human cancers ultimately result from events which take place in a single cell ... the origin of cancer from a single cell does not rule out influences on the rate of development of the cancer, influences which might be the result of tissue changes in an area surrounding the cell ... we do have evidence that some hormones can accelerate or retard the growth of certain cancers in the human being.
>
> — *John Gofman,* Radiation and Human Health, *p. 57, 1981.*

A PUBLIC-HEALTH issue arising with increasing frequency is the connection between a potentially harmful agent and an incidence of disease observed in an exposed community. We presently know very little about the influences exerted on human health and well-being by the many biologically active agents we release into the environment. Biological studies are intended, hopefully, to explain the fundamental processes by which these artificially introduced agents affect living organisms.

An aim of biological studies of radiation is to elucidate a *mechanism* — the successive biological reactions — that lead to a disease after irradiation. While a good deal has been learnt about the biological effects of ionising radiation, still little is known about the *pathogenesis* — the path of development of radiation-related disease. In the case of the more controversial health effects of non-ionising EMR, the connection between observed bioeffects and disease is even more elusive.

In this chapter some examples of frontier biological studies are described. They are illustrative of advanced research and thinking on radiation-related disease. But first, to understand how radiation interacts with our bodies, we need to know the rudiments of the basic unit of life — the living cell.

The living cell

A cell is the least bit of matter that we can say is 'alive'. There are billions of cells in the human body, almost all of them less than one-hundredth of a millimetre. Only cells can create other cells. They do this by dividing, each one into two daughter cells. Cells work in teams or colonies in a remarkably co-ordinated way, and they regulate cell division to replenish themselves only when needed by the organism.

Under a microscope we can see the cell's gross structure: a *nucleus* surrounded by a jelly-like *cytoplasm* contained in an outer cell *membrane* These cell parts are the ones that mostly feature in biological studies on radiation. There are other discernible parts. One is the *centriole*, which plays a key role in cell reproduction. Water makes up the greater part of all cells and accounts for about 80 per cent of their weight. Water molecules are a major target for radiation and, as stable as they usually are, ionising radiations can split them into very reactive parts (called free radicals) which are highly toxic to cells.

The outer membrane is so thin that it is difficult to identify as a distinctive part of the cell's structure. Nonetheless, a membrane is not just a thin inert casing; it plays a vital role in regulating the passage of food molecules and chemical messengers into and out of the cell. Electromagnetic radiation has been shown to affect the performance of membranes.

Cytoplasm is a more or less transparent jelly retained by the membrane. We get some idea of the cytoplasm's consistency from the white of an egg (a large cell) which is made viscous by the presence of the colloidal protein, albumin. Cytoplasm can be pictured as neither liquid nor solid, but made up of ordered arrays of giant enzyme and protein molecules like 'grains', invisible under the microscope. The cytoplasm's water molecules do not move randomly but are attached to the 'grains' in fixed arrangements. These molecular arrays pass chemical agents or electrical stimuli from one to the other, back and forth, along chains of command from membrane to nucleus.

The roughly spherical nucleus is the most visually evident part of a cell. The nucleus is pivotal to cell life and generally occupies a central position. Inside, the nucleus thread-like *chromosomes* store the genetic codes that guide the development and functioning of the living organism. Each body (*somatic*) cell has an identical set of chromosomes. The number of chromosomes in a cell varies with species; human somatic cells have 46 (see Figure 12.1).

The chromosomes

Cells are the basis of life; their chromosomes bestow on living creatures their distinctive forms and qualities. The genetic code

Figure 12.1: Human chromosomes in a dividing cell during the reproduction stage as they are seen under a microscope. The chromosomes are from a child with Down's syndrome — which results from a single extra chromosome indicated by the arrow

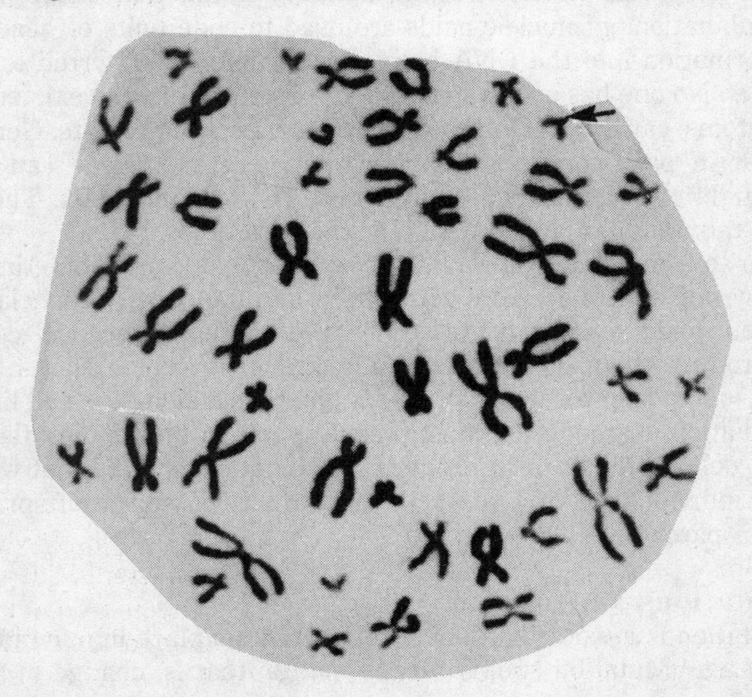

'written' into our chromosomes holds our priceless human heritage. Through our chromosomes we inherit our parents' characteristics: such as colour of eyes, skin complexion, physique, aptitudes. For the subtle blending of parental resemblances and mannerisms, forever a fascination to us, we have to thank the way that the 23 chromosomes in each of the male *sperm* and female *ova* pair off.

The genetic code is carried by long chain-like molecules of *deoxyribonucleic acid*, known as DNA. The DNA chain is formed by chemical bonds between small sugar and phosphate units and exists in a tightly coiled *double helix*. The genetic code is 'written' along this double-helix chain by varying the order in which four *nucleic acids* are paired between the sugar units along each chain. DNA controls the production of enzymes and proteins and the numerous functions of cytoplasm by sending out molecules of messenger chemicals.

The arrangement of the four nucleic acids along the DNA helix can be likened to the sets of dots and dashes used in Morse code to 'write' different letters of the alphabet. In much the same way, combinations of nucleic acids are used to code units of genetic information into the DNA helix. These units are referred to as *genes*. No one has seen a gene; their existence, like the existence of atoms, can only be inferred from the effects they create. Genes exist in pairs, one from each parent. and each gene can be thought of as a discrete section of the DNA double helix. There are thousands of genes within a single chromosome.

In so complex a system, it is no wonder that things sometimes go wrong and that errors get written into the code — especially when toxic agents intrude into the system. Because cells reproduce themselves so faithfully, any error which leaves the cell viable may be passed on to daughter cells along the cell line. Radiation is among the most potent agents in introducing flaws into cells. If the error is initiated in a somatic cell it dies out with the individual; if in a sex cell, the error is passed to offspring from parents.

Mutations

Mutation is a word that has become commonplace in our radiation age. Mutation simply means *change*; that is, change in the genetic codes of chromosomes. The mutations caused by radiation do not show up as unique bioeffects; there is nothing to distinguish mutations caused by radiations from those due to other causes.

Natural mutations arise from the malfunctions of a cell's reproductive processes. Biologists often disagree in their assessments of how much of a certain incidence of a disease can be attributed to radiation-induced mutations and how much to mutations induced by other environmental factors or natural causes.

One of the most frequent errors to occur during cell division is that, instead of 46 chromosomes going to each daughter cell, 45 go to one and 47 to the other. Chromosomes are normally matched in pairs, one from each parent. The daughter cell with one chromosome too many will have three of the same class and is *trisomic*, while the one lacking a chromosome is *monosomic*. One of the consequences of monosomy is mental retardation. Down's syndrome is one of the most common of the trisomic disorders.

Cells sometimes fail to complete their division into two daughter cells, and so their nucleus holds 92 instead of 46 chromosomes. Cells containing more than the normal complement of chromosomes are called *polyploid*. They occur with high frequency when cultivated white blood cells (leucocytes) are irradiated. Leucocytes from the Japanese crew of the fishing trawler *Lucky Dragon* were found to have four times the normal content of polyploid cells.

By far the greater number of mutations occur unseen in gene units. Gene mutation is considered to be due to a chemical rearrangement in the DNA chain. The toxic radicals formed when radiation interacts with the cytoplasm's water molecules are believed to be one of the main chemical agents in gene mutation. Mutant genes that are associated with the initiation of cancer are called *oncogenes*. It has been demonstrated that the same oncogene can cause more than one kind of cancer. A gene pair — one gene unit from each parent — controls a particular characteristic. In an individual, the effect of one gene in a pair may be very evident while the other may exert little control over determining the characteristic. These genes are respectively *dominant* and *recessive* genes.

The behaviour of dominant and recessive genes is readily illustrated with eye colour. The gene for brown eyes is dominant and for blue eyes recessive. If a mother's chromosome pair holds two blue-eyed genes then she will pass one of them on to her child. If the father is brown-eyed then he could have either a gene-pair both coded for brown, or one for brown and one for blue. In the former case there is no choice but for a dominant brown-eyed gene to be passed on to the child, who must have brown eyes. However, if the father has one blue-eyed gene in his pair then the possibility exists for it to pair up with the mother's blue-eyed gene. With two recessive genes in the pair, the child will be blue-eyed.

Dominant genes induced by radiation reveal themselves in the first generation. However, a recessive mutation will remain concealed — for several generations, perhaps — until it is matched with a similar recessive gene. The human cost of irradiation may not be counted for five, ten or more generations. Hundreds of genetic disorders can be attributed to either a dominant or recessive gene. For example, the distressing eye-disease retinoblastoma is caused by a dominant mutation. So too is dwarfism. Recessive genes are responsible for the blood disease sickle cell anaemia in black Africans, and the form of diabetes that results from failure of the pancreas to secrete insulin.

Many genetic defects are less severe and so are less evident and not easily identified with mutation. Mutations are believed to be responsible for a predisposition to degenerative diseases such as coronary diseases. It is now thought that mild mutations induced by environmental factors show up in later generations as an inherited predisposition to allergy, asthma, arthritis, hypertension and other debilities that undermine our health and well-being. It is possible that environmental pollution could have a compounding effect on health: mutations caused by a pollutant in one generation may reduce the immunity to another pollutant in a descendant.

The absorption of radiation energy

The transfer of radiant energy into living cells is an affair of the atoms. Radiations are emitted from atoms when electrons move and nuclei disintegrate. Biological changes occur when the atoms of living cells absorb the emitted radiant energy (see appendix: Where radiations come from).

Cells are composed of *molecules* formed by the chemical bonding of mostly carbon, hydrogen, oxygen, nitrogen, phosphorus and sulphur atoms. The 'housekeeping' of cells involves the making, breaking and rearranging of these chemical bonds. Cells are forever busy with chemical activity that requires a constant supply of energy normally provided by the *chemical* energy of food. However, radiation provides the life-giving energy for some cells as visible light does for the leaf cells of plants. Artificial non-ionising EMR can also give up its energy to cells; although, it seems, usually to their detriment.

The absorption of electromagnetic radiation by cells occurs as if the energy is delivered in 'packets' or *photons* (see box: Radiation energy is delivered in 'packets'). An X-ray or gamma-ray photon can deposit millions of times the energy of a radio-wave photon in the tissue it penetrates. An alpha or beta particle also packs a similar or more powerful punch of energy than do X-rays. High-energy photons are able to evict electrons from atoms to form *ions* whereas those from low-energy radio waves cannot. This capacity of photons to ionise, or not, gives us an energy threshold along the EMR spectrum above which radiations are ionising and below, non-ionising.

Compared to the energy required to break a *chemical bond* between say carbon and carbon, carbon and nitrogen or carbon and oxygen, the energy delivered by an X-ray photon is indeed enormous; the outcome can be likened to the havoc caused by a bull in a china shop. A direct hit by an X-ray photon on an electron in its atomic orbit sends the electron buzzing among other atoms, setting loose their fellow electrons. Many thousands of chemical bonds can be broken in this way by a single photon and the hurtling electrons that it sets loose. However, a photon usually does its damage among the molecules of one cell. The molecules themselves become excited in the process and run chemically amok. Excited enzyme, protein, sugar and other molecules change into aberrant molecules able to cause mutation and to upset the cell's metabolism.

There are differences in the way ionising photons transfer their energy to tissue. X-rays and gamma rays, like all EMR, travel at

RADIATION ENERGY IS DELIVERED IN 'PACKETS'

In 1899, after Heinrich Hertz had investigated his new *electromagnetic* radiation and showed it behaved in a manner similar to visible light, he confidently asserted that 'the wave theory of light is, from the point of view of human beings, a certainty'. Not for long. In 1905 Albert Einstein postulated that electromagnetic radiations behave as particles as well as waves, sometimes appearing as one and at other times the other.

Wave theory does not explain some radiation behaviour. When ultraviolet strikes a *negatively charged* zinc plate, electrons, the source of the negative charge, are knocked loose from the plate by the radiation. The plate loses its charge. When the zinc plate is exposed to *lower-frequency* visible light the electrons stay put, no matter how much the *intensity* of the light is increased. However, visible light removes electrons from the metal caesium.

After investigating the phenomena with a number of different metals it was found that each metal displays a specific *threshold* frequency. Increasing frequency above the threshold increased the *speed*, not the number, of the electrons knocked out. Higher frequency means higher energy. The phenomenon is known as the *photoelectric effect*.

Einstein explained the photoelectric effect as radiation behaving like a stream of tiny bullets. If a 'bullet' carried a certain minimum energy then it was capable of knocking an electron off the metal surface. If the energy was too low to evict an electron then increasing the number of 'bullets' would not help. Einstein developed his hypothesis from an earlier suggestion by a German physicist, Max Planck, that heated bodies emitted radiation in *quanta* or 'packets' and not continuously. He called a packet a *quantum* (Latin for quantity).

Einstein went a step further than Planck, and postulated that light has to be in 'packets' not only when it is emitted but when it is absorbed.

The energy of photons is measured in *electron-volts* (eV): that is, the energy gained by an electron when it falls through a potential of one volt. The energy of photons varies enormously. While RF photons carry only a small fraction of an electon-volt, X-rays and gamma rays carry millions of electron-volts (MeV).

the speed of light, giving them considerable penetrating power. Some of the rays pass through the target tissue without losing energy, which allows us to take photographs of internal organs. It also means that ionisation of atoms and molecules on the atomic scale occurs over a relatively long pathway. On the other hand, alpha and beta particles are electrically charged particles which travel at lower speeds and deposit their energy on the surface of tissues or soon after entering. The alpha particle has been described as an atomic cannonball with short range but one which is devastating to the tissue at the spot where it lodges.

Though ionising photons and particles can be enormously destructive, it may come as a surprise that in ordinary terms they carry precious little energy. The havoc wreaked by photons on the molecular architecture of cells is not because of the *quantity* but rather the *quality* of their energy packet. The actual quantity of energy absorbed by the whole human body from a lethal dose of ionising radiation amounts to less than 100 calories, or no more than the heat energy in a single bite of hot food. The reason for the devastating effect of ionising photons lies in the way the *absorbed* energy is so highly concentrated that it can be imparted to just a single atom. In the case of a mouthful of hot food, the energy is distributed over millions of atoms.

The ultraviolet photons, which are the next step down in energy level from X-rays, are not energetic enough to knock electrons out of atoms. But they can excite molecules enough to break their chemical bonds and bring about chemical change. Our suntan results from ultraviolet photons initiating chemical change that produces skin pigments.

Visible light photons can also initiate chemical change. Chlorophyll in cells uses visible light to convert carbon dioxide and water into sugars — a process which life on earth depends upon for its existence.

Below visible light, further down the energy scale, are infrared and RF. These may excite molecules to vibrate but not break the *chemical bonds* in cell molecules. Here we have another energy threshold below which bioeffects may be caused by heat generated by molecular vibrations if the radiation intensity is high enough. However, the bioeffects of ELF radiation, at the bottom of the EMR energy scale, cannot be explained by the transfer of photon energy, but must be explained in other ways. One school of thought is that biological systems are sensitive to the frequencies of pulsed or modulated *waveforms*.

Radiation and the cancer cell

'What we did was to expose human colon cancer cells to four field conditions — electrical, magnetic, combined electrical-magnetic, and a fourth with no fields (to act as a control) for 24 hours, and then remove the cells from the chamber', Dr Jerry Phillips explained at a student seminar in socio-environmental studies. 'We counted the cells. We wanted to see whether or not the exposure itself had any effect on the cells. Did it kill the cells? Did the cells simply drop dead on their own? What we found was that if we put in a good viable suspension of cells then we pulled out the same good viable suspension of cells. The exposure neither decreased nor increased the death of the cells.'

The *cell culture* is one which Dr Phillips's research institute uses to assay cancer drugs. 'We know these cells do not necessarily behave the same on a daily basis any more than people do', he said, and so the exposures were repeated with many other samples.

The cells are capable of *cloning* and producing colonies. Samples of the irradiated cells were stirred in a nutrient solution in glass dishes to isolate them as single cells and then incubated for seven days. 'Basically what we found in our experiments', Dr Phillips reported, 'was this: cells that were exposed to electromagnetic fields produced a greater number of *colonies* than the control cells that received no exposure to electric or magnetic fields or a combination of those fields'.[1]

Cell membranes contain proteins that are characteristic of the cell. A function of the proteins is to *recognise* other cells and molecules in the environment which they have either to interact with or to reject. Phillips looked at how this recognition capacity might be altered in cancer cells by electric and magnetic fields. Of particular interest was the recognition by lymphocytes (natural killer cells), one of the body's defences against foreign cells. So, exposed and control cells were mixed with killer cells in separate dishes and incubated for up to several months. It was found that a certain percentage of control cells were destroyed by killer cells while cells exposed to electromagnetic fields had developed a measurable resistance.

Jerry Phillips had been invited to Melbourne by the Collingwood Residents Association to advise them on recent biological research into non-ionising EMR and the implications for human health. The issue of whether or not something in the environment poses a threat to health is a difficult question at best; admittedly, there are severe limits to how findings on cell cultures can be related to a real-life problem. They have to be appraised along with all other relevant research. 'It is data like I have shown you today on which the assessment has in part to be based. It is also the epidemiological data and the results of a large number of other studies.'

In another laboratory the results from exposing colon cancer cells to ELF radiation were equivocal.[2] Yet in still another laboratory where lymphocytes were targeted against lymphoma cells (lymph cancer), microwaves modulated at low frequencies significantly reduced cell destruction. Plain unmodulated carrier waves had no effect.[3]

The studies point to modulation at low frequency as a factor in the observed effects of non-ionising EMR on the immune processes. Guy's findings of tumour growth in the adrenal glands of rats (described in chapter 3) give added support to the hypothesis that biological activity is associated with modulation. Other evidence comes from the way in which pulsed magnetic fields affect the secretion rate of noradrenalin, a nerve hormone (neurotransmitter), from the adrenal gland.[4]

'In the assessment of EMR and similar risks, what you have to keep in mind', Jerry Phillips told his student audience, 'is that there is no single opinion that you will ever find relative to any issue involving human health. Look at the problems of asbestos; look at the problems with cigarette smoking . . . You should at least use the data that show biological systems to be at risk to exercise prudence in exposing the general public to electromagnetic fields'.

Cell communication systems

'Recent observations have opened doors to completely new concepts of communication between cells as they "whisper together" across barriers of cell membranes', Dr Ross Adey said during the 1986 Abbie memorial lecture in Adelaide.[5] Natural electromagnetism is produced by the stimulation of the membrane surface by such molecules as hormones, enzymes, anti-bodies and neurotransmitters, and is used by cells for communication. The signals from artificial EMR may intervene. The cell membrane, according to Dr Adey, 'is turning out to be a very sensitive detector and a very sensitive amplifier of a very wide range of fields — the low-frequency fields of the power-line type on the one hand and radio and microwave fields which are modulated at low frequencies on the other'.[6]

In the late 1960s research showed that ELF waves at 16Hz altered the behaviour of humans and monkeys.[7] In later work Adey's research team observed changes in the electrical activity of the brains of cats, recorded on *electroencephalograms*, when exposed to ELF waves. From these results it was deduced that nerve tissue was involved, and so it was decided to measure changes in calcium flow in brain tissue — a phenomenon associated with nerve function. ELF waves, or RF waves modulated by ELF waves, disturbed the calcium balance in brain cell membranes.[8] The team reported an effect on the flow of calcium ions, but they found that it did not rise consistently with increases in either frequency or intensity. Instead the effects appeared only at certain frequencies and intensities which they described as 'windows' in the EMR spectrum, to which cell membranes could make specific responses.

We know little about the biological events that occur over the time — up to 40 years — between when a cancer is initiated and its clinical diagnosis. It is generally agreed that cancers develop in two stages. The first stage is a mutation in a *stem* cell — a cell that replenishes body cells. The cell acquires a potential to reproduce without the restraints applying to normal cell growth. This seed cell and possibly a small number of descendant cells may remain dormant for years before beginning rapid malignant growth. Only in the case of leukaemia can adult cancers be observed clinically less than ten years after initiation.

It is during the second, or *promotion*, stage of a cancer that growth control mechanisms are overcome. The indications are that this is brought about by such factors as ageing, loss of immune response, changes in hormone secretions, a biologically active agent or infection. Only when one of these factors comes into play does the mutant cell go into its malignant phase.

It is speculated that the observed induction of cancer by non-ionising EMR occurs during the promotion stage. The known ability of EMR to alter the rate of hormone secretions suggests itself as a possible mechanism for EMR promotion of cancer. People have to be exposed to EMR over a long period before the health effects become evident. 'This first came to light in some of the studies that Nancy Wertheimer did in Denver,' Ross Adey said, 'where she had at least suggestive evidence that there is a minimum period of three years after exposure to high-voltage power lines before the threshold is crossed. Thereafter there may be quite a long period before you are safe even if you withdraw yourself from that exposure.'

There is other evidence for a latent period after initiation before a cancer manifests itself. For instance, if you are a smoker you don't get lung cancer from a brief period of smoking. Also, after you stop smoking you continue to be at risk for a period; eventually, that risk appears to decline back to somewhere near baseline level.[9]

Radiation injury to chromosomes

Observed under a microscope, the physical damage to chromosomes from ionising radiation is similar to abnormalities observed in chromosomes of cancer cells. This similarity may be taken as a fair indication that ionising radiation can cause cancer by chromosome injury.

In the 1950s two discoveries greatly enhanced the scope for investigating chromosome abnormality and injury. One was the discovery of *inhibitors* that cause cells to pause in their division at a point where the arms of chromosomes can be examined under the optical microscope for length and shape.

The other was that lowering the salt content swelled the chromosomes, causing them to separate into discrete units (see Figure 12.1). This meant that chromosomes could be counted accurately, their pairs matched and abnormalities in their shapes noted. Before these techniques were available microscopic images could not be produced. Until then, images of stained chromosomes were so fuzzy that the number in human cells were thought to be 48 and not 46, which it is now known to be.

With the use of these new techniques, Nowell and Hungerford in 1960 demonstrated for the first time a link between a chromosome abnormality seen under a microscope and chronic myelogenous leukaemia. This form of cancer can result from radiation acting on bone marrow. The researchers found that in patients suffering this disease an arm of the G-22 chromosome was unusually shortened. This could only mean that at some stage a part of the arm had suffered a *deletion*.

Myelogenous leukaemia can remain mild for years but at some stage becomes acute and terminal. In their study of the acute stage, researchers found that the fragments excised from chromosome G-22 had *translocated* onto the G-9 chromosome. It was also found during the acute phase of the disease that the chromosomes developed an *isochromosome* from a previously normal E-17 chromosome.[10] An isochrome occurs during cell division when the centromere (chromosome pair junction) divides across, instead of along, the length of a chromosome pair. The finding that this chromosome alteration occurred coincidentally with a change from a mild to acute cancer is evidence that chromosomal alterations can cause malignancy.

FORTY years after the atomic bomb explosion at Hiroshima, people exposed to its radiations still have abnormal chromosomes in their bone-marrow cells. The frequency of the abnormal chromosomes is a measure of their radiation dose. The same relationship between radiation dose and chromosome alterations is found in irradiated cultivated tissues.[11] Of course, there must be many physical changes in chromosomes brought about by radiation that cannot be observed microscopically but which would lead to cancer or other disease. Neither can internal dislocations in the DNA helix be observed under a microscope.

The Lawrence Livermore laboratory in California has developed a technique by which gene damage may be measured. Red blood cells (erythrocytes) have two variants of the protein glycophorin present in their membrane. Two separate genes code for the two protein variants. One gene is much more radiation sensitive than the other. By labelling the two protein variants with dyes it is possible to measure the relative changes in their rate of production and so record the extent of gene mutation. It was found possible using this method to distinguish between a group of people that had undergone treatments with medical radiation (and chemical mutagens) and groups that had not. When the test was applied to survivors of Hiroshima and Chernobyl it was possible to obtain some measure of the ionising radiation dose individuals had received.[12]

Genetic inheritance

Nature puts up barriers to protect future generations from inheriting mutations. Cells carrying mutations frequently die out because they fail to complete cell division. Cells with abnormal genes may not reproduce because the genes cause infertility or bring on miscarriage from failure of the embryo or foetus to develop. Cells have repair mechanisms to mend their damaged chromosomes. Many deleted chromosome fragments are restored to their proper place by these repair mechanisms. However, microscopic examination cannot tell us how well any particular repair is done.

Nature's barriers do not always block faulty genes and chromosomes which, unfortunately, can slip through to later generations. The scope for damage by ionising radiations depends on where the mutated cells are located. A mutation in a somatic cell can only affect the organ with the damaged cell; it may cause malignancy — which then spreads to other areas — but it can go no further than the individual. When the mutation occurs in the sperm or ovum the offspring may inherit a gene or chromosome defect. The parent may have been irradiated only days before conception, or it could have been an ancestor affected centuries ago.

Mutational effects carried through the reproductive cycle are referred to as *genetic*. Irradiation of the embryo or foetus, that is *in utero*, can lead to *congenital* or *teratogenic* effects. In utero effects are often classified as genetic effects. However, unless the gonads of the offspring are involved, the defects are not passed on to following generations.

Evidence of genetic effects of an ionising radiation first came from laboratory studies in 1926-27 when an American geneticist,

Herman Muller, irradiated *Drosophila* (fruit fly) with X-rays.[13] Visible mutational effects were observed down to very low doses of the radiation, and the same result was obtained whether the dose was given all at once or was spread over a period of time. In 1946, he received the Nobel Prize for Medicine.

Muller's findings of over 60 years ago have not been given the prominence they deserve. His findings should be worrying us for two very basic reasons. One is that no dose of an ionising radiation is without some genetic effect, yet we are exposing ourselves to increasing levels of this type of radiation from medical treatment and nuclear activities. The other reason is that the genetic consequences of exposure are cumulative. Mildly detrimental mutations accumulate in the human genetic bank; this can lead to a deterioration of life qualities in future generations.

Biological time-giver

Possibly the most consistently demonstrated effect of non-ionising EMR has been the change in the behavioural characteristics of animals.

The rhythm of our daily life, when we feel sleepy, active, hungry or whatever, is thought to be directly linked with the diurnal cycle of alternating night and day. Other creatures follow similar behavioural rhythms, although those leading nocturnal lives like the owl are out of phase with us. This diurnal variation of body functions is called the *circadian rhythm*. There can be upsets. After an aeroplane lands us where the night-and-day cycle is out of phase with that from whence we came, we suffer jet lag.

However, biological studies are showing that circadian rhythm goes deeper than just our outward behaviour. The rhythm is co-ordinated by an interaction of *in phase* internal physiological and biochemical cycles. Besides these internal co-ordinators there are external ones. They are called *zeitgebers* or 'time-givers', and play a part in synchronising the highs and lows in the rhythm of our daily activities.

Natural electromagnetic fields are thought to be among the zeitgebers. Changes in the natural fields, or the intrusion of artificial ones, have been found to cause shifts in circadian rhythm that are related to biochemical reactions of the endocrine and nerve systems. It is these two systems that provide the vital communications network of the body. The nerves act like a telegraphic wire service, while the endocrine glands function like a radio hook-up, putting out messages into the bloodstream in the form of highly specialised chemical substances called hormones that elicit a response in an organ. Changes in the biochemical cycles in these two communications systems manifest themselves in shifts in behavioural rhythms. Outwardly, a shift may show up in a lessening of our readiness to concentrate: in a classroom, this might make a pupil listless and inattentive. In the workplace it can leave individuals more vulnerable than usual to accident. Seasonal shifts have been related to psychological depression and changes in the immune system.

In the 1960s, Dr Rutger Wever built two underground bunkers. In both bunkers people were insulated from the usual time-givers: changes in light, temperature and sound. One bunker was shielded from electromagnetic fields while the other was not. Wever observed several hundred people over periods of up to two months, plotting their body temperature, sleep-awake cycle and urinary excretion of sodium, potassium and calcium — elements that take part in the functioning of the nerve system and brain tissue.

People in both bunkers experienced shifts in their circadian rhythm; but those in the shielded bunker had significantly larger shifts. The circadian rhythm of those in contact with the earth's electromagnetic field remained close to 24 hours, but the circadian rhythm of those people shielded from the field was desynchronised. Their biochemical cycles became completely out of phase with each other. When Wever introduced an electromagnetic field of 10 Hz the normal rhythms were restored. This is the dominant frequency of brain waves recorded on electroencephalograms and in the earth's magnetic field.[14]

WE have all wondered about the marvellous migratory behaviour of birds. Mutton birds, for example, migrate each autumn from Australia to Siberia, and return to their southern habitat the next spring. The birds, after breeding on the Bass Strait islands, fly off on their long northward journey over thousands of kilometres of Pacific waters. The puzzle has been how birds can navigate so precisely as to find their destination unfailingly. After long and elaborate investigations it seems that migratory birds use a number of cues. One cue is the earth's magnetic field; and the evidence is convincing that the primary response to the field occurs in the pineal organ. Physical changes have been observed in the gland's cells, according to the orientation of a pigeon's head to the magnetic field.[15]

The tiny pineal organ located in the centre of the cranium produces the enzymes that catalyse the synthesis of melatonin and seretonin. Through the secretion of these hormones, the organ exerts ubiquitous control over the gonads, thymus, adrenal gland and the thyroid. In turn, the rate of melatonin secretion helps to regulate the circadian rhythm. An increase in melatonin extends the duration of sleep or decreases an individual's attentiveness. The onset of puberty is believed to be associated with a sudden decrease in the night-time secretion of melatonin.

Weak electric and magnetic fields affect the pineal function. When mice were pre-exposed to a weak magnetic field for ten days before morphine injections they became tolerant to the drug, whereas when exposed to a sham field they became sensitive.[16] In the cerebral spinal fluid of monkeys, seratonin remained depressed; the suppression of dopamine, which influences blood pressure, was reversed.[17] This indicates that bioeffects of ELF radiation may produce more permanent effects than just transient alterations in behaviour associated with disruptions to the circadian rhythm.

Chapter 13

Where the experts differ

> The modelling of radiation risks can provide only a loose and flexible framework within which *political* debate over nuclear policies can take place. The most important question is then no longer "what is a precise risk?", but "who benefits from the doubt?". This is a problem of accountability and control.
>
> — *David Crouch, 'The Role of Predictive Modelling: Social and Scientific Problems of Radiation Risk Assessment',* Radiation and Health, *pp. 57-8, 1987.*

> Misreporting or distorting the scientific record at hearings does tend to alienate journalists and the public, but is not of itself an insurmountable problem. Others can and have looked at the scientific record and can and have drawn their own conclusions.
>
> — *Allan Frey, 'From the laboratory to the court-room, science, scientists and the regulatory process',* Risk/Benefit Analysis — The Microwave Case, *p. 208, 1982.*

A COMMUNITY which has been exposed to a radiation source and which has the temerity to question the official line — that 'there's nothing to worry about' — may be taken aback by the way it is publicly admonished by radiation authorities. Yet community concerns will almost certainly have been aroused by scientists with reputations in the relevant field. The authorities would like us to think their radiation health regulations are based on undisputed scientific evidence. In fact, experts differ widely about the dangers posed to human health by exposure to radiation.

But where does a community perceiving a radiation health

problem begin to come to grips with the range of scientific opinion? A power-line action group began its battle with the Victorian health department by collecting a pile of photocopied review articles and research papers on related biological and health studies. Review articles and books by specialists in the field proved especially helpful. The group had first sought advice from scientists around the world on the source-material to consult. They documented the scientific basis of the action group's concerns about the environmental and health effects of the proposed power line.

A public benefit of this community action was the airing of the whole health issue — not just a narrow band of expert opinion which the authorities wanted people to hear. As well, Victoria's political agenda now included the principle that a community had a right to be fully informed about the likely adverse health effects of a technological installation planned for its area.

Of course, political debate on radiation safety is not new. It began in the 1950s with the exposure of populations to ionising radiation emitted by radioactive fallout from nuclear weapons testing. Debate on the safety of non-ionising EMR is more recent. Differences of scientific opinion are being heard with increasing frequency in public debate on radiation safety. This chapter outlines the scientific basis of some of these differences. Some topics are specific to either ionising or non-ionising radiations. However, two key topics that overarch the whole radiation debate are the *connection* between radiation exposure and disease, and the precise nature of the *risk*.

Making a connection

Where there is reason to suspect that a radiation source is causing adverse health effects (for instance, cancer from a nuclear plant or miscarriages from VDTs), the evidence of an association between the two may be sought simply by comparing the exposed population with a 'matched' unexposed population. Finding an 'excess' number of disease cases in the exposed population is suggestive of the disease being connected with the radiation exposure. The credibility of the epidemiological data rests on two considerations: how rigorously the two populations are matched and whether other environmental factors can reasonably be ruled out.

A looser statistical approach is to correlate the number of deaths in an exposed population with the 'national norm'. The 'observed' and 'expected' number of deaths for an exposed population are determined, to see if there is an excess in the exposed population which is too large to be attributed to chance when tested by the rules of statistics. An example of this approach is a recent analysis of US national mortality data by Jay Gould, a leading US geneticist, which was suggestive of low-level radiation being responsible for a deterioration of the human immune system. In a recent book, *Deadly Deceit*, he and Benjamin Goldman reproduced graphs to illustrate a connection between fallout radionuclides from nuclear plants (including fallout in the United States from Chernobyl) and excess mortality among the very young and old [whose immune systems are either not fully developed or are deteriorating] and the time, place and activity of fallout. Statistical correlations of an incidence of disease and radiation, such as this, do not *prove* a connection between the two; however, Gould suggests, they do provide clues of where to look for the source of an outbreak of disease.

A more precise way of demonstrating a connection between a radiation source and an incidence of disease in an exposed population is to derive a *dosimetry* from determinations of the *doses* to individuals and the *collective* dose to the population (see box: Ionising radiation doses to individuals and populations). Having estimated the radiation dose, the 'expected' number of deaths can be obtained by reference to the radiation risk-estimates compiled from past studies of exposed populations (see Table 13.1). A reasonably close correspondence between the 'observed' and the 'expected' number of disease cases is considered to be a positive finding of a *causal link* between the radiation source and the incidence of disease. However, the opportunities to gather the data for the dosimetry of long-term exposure of populations to radiation sources are limited. And those responsible for fallout do not have any great enthusiasm for keeping consistent records of their releases into the environment. Epidemiologist Joseph Lyons gathered data to assist the 'downwinders' living in Utah in their battle for compensation from the USAEC and the US government for fallout from nuclear weapons tests in the Nevada desert. 'While dosimetry is important', Lyons said, 'you cannot conclude that there are no adverse health effects based on dosimetric study . . . The appropriate method of determining health effects is to *measure the health of the population*, not to guess at the level of exposure'.[1]

Wertheimer derived a dosimetry for ELF magnetic fields experienced by householders living alongside power lines. However, measuring doses of non-ionising EMR is generally only possible in laboratory studies on animals; and the data obtained is not readily related to human exposure. Dosimetry of non-ionising EMR is also complicated by the biological effects occurring – if the Adey hypothesis is valid – in frequency and intensity 'windows'.

But even estimating the level of dose of an ionising radiation to a population – for example the Seascale community – is fraught with difficulty. Much is speculative. In order to estimate

Table 13.1: Estimates of a lifetime risk of fatal cancer induced by ionising radiations

		Cancer deaths 100 000 exposed to 10 milliSv
International Commission on Radiological Protection (ICRP)	1977	12
	1990	48-60
United Nations Scientific Committee on the Effects of Atomic Radiation (UNSCEAR)	1977	10
John Gofman (*Radiation and Health*)	1981	377
Rosalie Bertell (*Handbook for Estimating Health Effects of Ionising Radiation*)	1982	55-165
US Biological Effects of Ionising Radiation Committee (BEIR V)	1990	80

the collective dose to an exposed population, computer models are used to simulate migration of deposited radionuclides through the environment and their metabolic behaviour in the human body. The design of the models and personal discretion exercised over what data is fed into the computer (and guesstimates where data is lacking) leave plenty of scope for experts to differ. Once radionuclides enter the environment they may be physically transported by wind and water, or else they may be transferred by innumerable biochemical processes operating in ecosystems. Eventually they are transferred along food chains to humans. The rate at which this happens depends, among other things, on soil type. The strontium-90 in fallout from Chernobyl found its way into farm products at vastly different rates according to the soil on which it was deposited.

Figure 13.1: The environmental pathways along which radionuclides find their way into our food. Such diagrams are used as the basis for computer models to calculate doses of radionuclides deposited by nuclear activities. They reach humans through the food they eat or accumulate in the deep layers of soil. The models are a great simplification of the natural processes and are being constantly revised

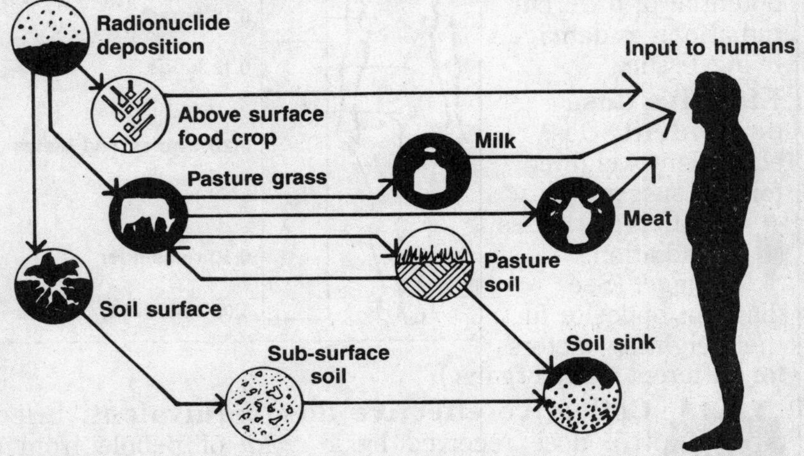

Source: *Radiation: Doses, Effects, Risks*, United Nations Environment Program.

Radionuclides accumulate in milk, meat, fish, cereals and other common foodstuffs; and the contribution to the radiation dose from each food item in the average diet (not exceptional tastes) must be estimated. Some ionising radiations cause more biological damage than others. The dose must be adjusted according to whether the radiation is alpha, beta or gamma. Alpha rays (for example, from inhaled plutonium) are considered to be twenty times more damaging than either beta or gamma rays. Some parts of the body are more susceptible to radiation damage than others; for example, a dose of radiation to the lungs is more likely to cause a *fatal* cancer than the same dose to the thyroid. The dose to each organ is 'weighted' to take account of specific radiation sensitivity. Radionuclides differ in their rates of decay and also in the rate at which they are excreted from the

IONISING RADIATION DOSES TO INDIVIDUALS AND POPULATIONS

Absorbed dose: The dose imparted by radiation to living tissue.

Dose equivalent: The absorbed dose weighted for the potential of different radiations to damage living tissue.

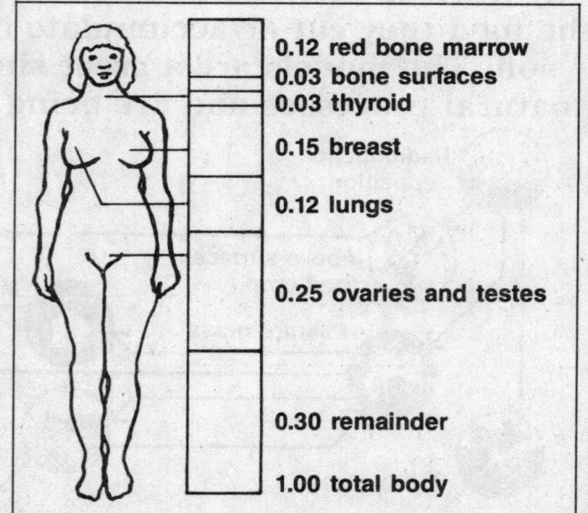

Effective dose equivalent: Dose equivalent weighted for the susceptibility of different tissues to radiation damage. (See diagram opposite for the weighting factors for different body organs.)

Collective effective dose equivalent: Effective dose received by a *group* of people from a source of radiation.

body; estimates of biological damage to body tissue must take account of both these variables.

From time to time the models are changed to accommodate new data. The models used by radiation advisory bodies for the dosimetry of the artificial transuranic elements — according to some critics — are crude, grossly underestimate the irradiation of internal body tissues, and are liable to substantial revision as more research is done.

Risk

International and national radiation advisory bodies recommend radiation safety standards in which they set limits to public and occupational exposures. Two quite separate processes are involved in deciding these limits: one is a *quantitative* assessment

Collective effective dose equivalent commitment: Collective effective dose equivalent delivered over time to *generations* of people.

Potential to damage tissue:

Alpha rays are positively charged particles with low penetrating power but are damaging when emitted inside the body. They are 20 times more damaging to tissue than beta or gamma rays.

Beta rays are negatively charged particles with low penetrating power and are most damaging when emitted inside the body.

Gamma rays are electromagnetic radiation with the power to penetrate the body from outside sources.

Units:

Gray is the unit of the absorbed dose: one joule imparted to 1kg of tissue.

Sievert is the unit of the dose equivalent. This is the absorbed dose weighted according to the potential of the radiation to do damage. One sievert also corresponds to one joule/kilogram

Becquerel is a unit of radioactivity. One becquerel unit corresponds to one disintegration a second of any radionuclide.

of the risk of radiation exposure; the other is deciding upon a 'socially acceptable' level of risk. The former has a legitimate scientific basis. The latter relies on unspoken value judgements by the experts about a balance between the economic and social 'benefits' of the activities causing radiation emissions and the 'costs' to public health and the environment; it has no genuine scientific basis and should not be left to the experts.

Quantitative assessment of the risk of exposure to ionising radiations is based largely on studies of the incidence of cancer in populations exposed in the past (for instance, the Life Span Study on the Hibakusha in Hiroshima and Nagasaki, and the Mancuso study on workers at the Hanford nuclear plant). From these studies *dose-response* relationships have been developed (see Figure 13.2) that give the number of cancers *proportional* to the dose absorbed by a population (see Table 13.1). Estimates of the numbers of cancers can be seen to vary over a wide range according to who did the calculation. A lot depends on the selection of the data. The ICRP, for instance, excludes findings where a precise dosimetry cannot be derived; it relies mostly on the Life Span Study and interprets its data conservatively. Some risk estimates are determined for a shorter time-span after irradiation than others. Selective use of data can exclude studies that taken together are supportive of certain trends, but considered separately offer only loose indications of risk. Despite these differences, risk estimates allow for genuine scientific debate and provide a rough yardstick useful for communities to gauge radiation risks.

RISK ESTIMATION

Risk may be expressed as 'so many per a certain sized group' such as '40 per 100 000'. Risk may also be expressed as 'one chance in so many' such as '1 chance in 250'. To convert a risk of 40 per 100 000 into the odds, divide 100 000 by 40 to get 1 chance in 250. To convert these odds into the number in a group of 100 000, divide 100 000 by 250 to get 40 per 100 000.

Use of *comparative* risk is a different kettle of fish. Nuclear industry publications offer the most unlikely risk comparisons —

between, say radiation exposure from nuclear activities and so many X-ray exams, being struck by lightning, or being in an aeroplane crash.

Risk analysts assess the chance of fatality for every conceivable human activity from smoking, to rock climbing, or crossing the road (see Table 13.2). The numbers obtained are used to determine a 'socially acceptable' risk of fatality, which radiation advisory bodies use to decide the level of risk embodied in their radiation safety standards.[2] It is inadmissable to compare unlike risks in this way. There is a great difference in the way people perceive voluntary and imposed risks. An individual may accept the risk of an X-ray and yet feel justified in resisting a risk imposed by the operators of a nuclear plant or a radar transmitter. People who engage in risky sports may resent the imposition of a risk they feel helpless to do anything about.

Protagonists of radiation-related projects feel irritated by what they see as irrational behaviour. But is it so irrational that people should want to decide the risks they take with their own lives? People rightly judge imposed risks in the light of the benefits

Figure 13.2: The various models used by epidemiologists to describe the dose-response relationship of ionising radiations at low doses of radiation. [a] linear; [b] linear-thresholds; [c] linear-quadratic; [d] supralinear

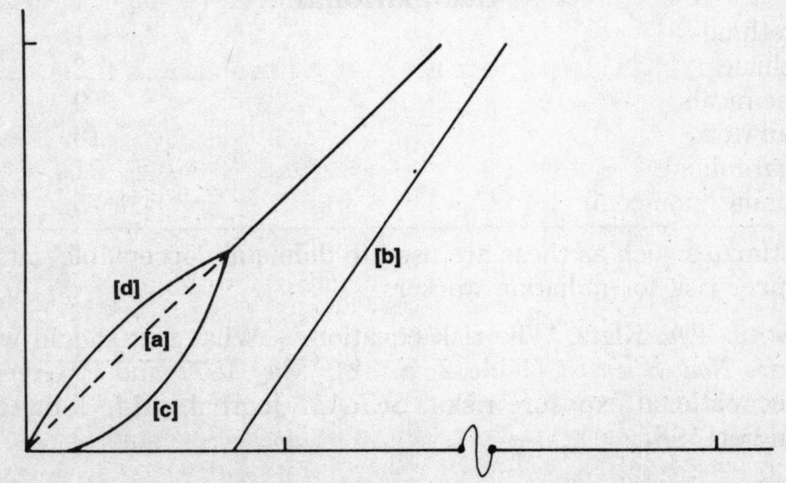

they enjoy, and they may feel justified in demanding alternative technologies that present more palatable risks.

A radiation risk may be weighed in terms of an individual's chance of contracting a disease. On the other hand, a risk may be looked at in social terms as a consequence for public health. The promotors and controllers of radiation sources like to talk about the 'very, very small risk' to individuals. Those concerned about environmental health see the risk in terms of hundreds or

Table 13.2: Voluntary, involuntary and occupational risks

Activity	Deaths per 100 000 per annum
Voluntary	
Taking a contraceptive pill	2
Football	4
Rock climbing	14
Smoking	500
Involuntary	
Run over on road	6
Leukaemia	8
Influenza	20
Occupational	
Clothing	1
Vehicle	2
Chemicals	9
Railways	18
Coal mining	21
Mining (non-coal)	75

Estimates such as these are used to define an 'acceptable' fatal cancer risk for radiation workers.

Source: T.A. Kletz, 'The risk equation — What risk should we run?', *New Scientist*, Table 2, p. 321, May 1977 and P. Green 'Occupational exposure risks', *SCRAM* Journal, p.14, Table 1, Jan/Feb 1987.

possibly thousands of people being affected and many hospitalised during their lifetime as a result of radiation exposure.

The chance of a ten-year-old boy getting cancer in later life from a full-mouth dental X-ray exam has been estimated to be one chance in 600. That is, in a population where such an exam is common practice, one boy in every 600, in later life, will probably contract a thyroid or other cancer. Even for the individual, this is a risk that should be weighed carefully against the likely benefit. But from a public health viewpoint the consequences are also significant. For Australia, where this exam is common, the aggregate must amount to many cancers over the whole population.

In some situations the risk continues after the activity has finished; for other activities there is no continuing risk. Thus there is no risk of dying from an aircraft crash once you have landed safely. However, the consequences of exposure to radiation for even a short time represent a risk spread over a long timespan. It can mean suffering and anguish in a family over many years.

The 'scientific' determination of what are 'socially acceptable' risks by expert committees is being used by governments to regulate hazards in the workplace and environment. It is a process by which health policy is decided by expert committees, rather than more properly through social and political processes.

IONISING RADIATIONS

The dose-reponse relationship

The accepted relationship between radiation dose and excess cancer — the dose response — is represented by the *linear model* depicted by the straight line [a] in Figure 13.2. If a threshold existed then it would be represented by line [b], where the excess of cancers reaches zero before the radiation dose.

In the linear model, depicted by line [a], the health risk *per radiation unit* is the same at lower dose levels as it is at higher dose levels. Thus, if each member of a population of 1 000 receives 100 units of radiation and 10 persons die prematurely of radiation-related disease, it is valid to say that in a population of

100 000, in which each member receives 1 unit, 10 will also die prematurely.

Two other models are put forward as alternatives to the linear model. One is the *linear-quadratic* model. It is depicted by line [c] showing a proportionately *lower* risk at low doses. The other is the *supralinear* model depicted by line [d]. Here the risk is proportionately *higher* at low doses. There is growing evidence to support this latter model.[3]

By the late 1970s, and after years of argument, the ICRP and most national radiation protection bodies had accepted the linear model as the basis for recommending 'acceptable' limits of radiation exposure. However, official pronouncements often convey the idea that small amounts of radiation are harmless. This is a tacit reinstatement of the 'safe dose' idea; if each increment of radiation is harmless then so too is their cumulative dose. Equally wrong is Australian customs practice of allowing an importer to mix food contaminated with radionuclides with uncontaminated food in order to bring it within the regulation radiation count. The consequences for public health are the same.[4]

The concept of 'safe dose' has also been questioned in other ways. It is suggested that taking the dose-response line [a] in Figure 13.2 down to zero relies on either imprecise epidemiological surveys at low doses (100 milliSv down to zero) or doubtful extrapolations from high doses. If so, what is a 'safe dose' in the light of generally accepted findings? How close to zero?

Israeli children, whose scalps were given a dose of 75 milliSv during treatment for ringworm, suffered an excess of thyroid cancer.[5] Radford reported that revisions of the Hiroshima and Nagasaki data showed the dose-response relationship pretty close to a straight line down to 30 milliSv.[6] The more controversial Mancuso findings on Hanford workers showed excess cancers were down to 10 milliSv.[7] In fact, convincing evidence against a 'safe dose' has been with us since 1970, when Stewart's Oxford study of child cancer showed a linear dose-response down to 2.5milliSv for the foetus.[8]

Cell repair-mechanisms have been suggested as a way the body can recover from low radiation doses. If this is so, a series of low doses over a long period would be less harmful than an equivalent single dose being absorbed at one time.

As the lowest dose-rates proved to cause excess cancer, the chance of any one cell being hit more than once can be calculated to be very low. The transfer of a photon's energy to a cell can be considered as an all-or-nothing affair. This situation — where any single cell is almost always injured by only one photon, and each cell's repair-mechanism either copes or fails — suggests that there is the same risk from each photon, whether it comes amidst a downpour or a light shower. In other words, each damaged cell has an equal chance of mutating whether the dose is low or high.[9] Since cancer is observed at low doses the repair-mechanism of some cells clearly does fail.

The lack of evidence of an association between *natural* radiation and cancer is sometimes offered as evidence that repair-mechanisms can cope at least with constant exposure to one to two milliGy. Alice Stewart recently compared variations in background gamma radiation with her Oxford-survey data on child cancer in Britain. Stewart estimated that 80 per cent of all cancer deaths before the age of 16 are caused by either prenatal X-rays or natural gamma radiation, with the latter being the major source of cancer.[10]

Permissible radiation doses

From time to time the ICRP reviews the data on risk-estimates and recommends a *permissible* occupational dose of ionising radiation. Most national radiation protection bodies, including the radiation committee of the NHMRC, adopt ICRP recommendations. Since the inception of ionising-radiation standards the trend has been downwards drastically. In 1950 the dose-limit was reduced from 300 to 150 milliSv per annum. It was reduced again, in 1956, to 50 milliSv (Figure 13.3). In 1977 the ICRP reaffirmed the 50 milliSv limit. In 1985 the recommended public dose-limit was reduced from five to one milliSv per annum, except in an emergency.

The recommended occupational limit is not meant to signify a 'safe' radiation dose; rather, it represents a boundary above which the dose is unacceptably dangerous. According to the ICRP, radiation doses should be kept 'as low as reasonably achievable . . . economic and social factors being taken into account'. The ICRP, in its 1977 recommendation, suggested that

Figure 13.3: Over the years the permissible dose of ionising radiation has been revised drastically downwards. The International Commission on Radiological Protection started recommending permissible exposures in 1934

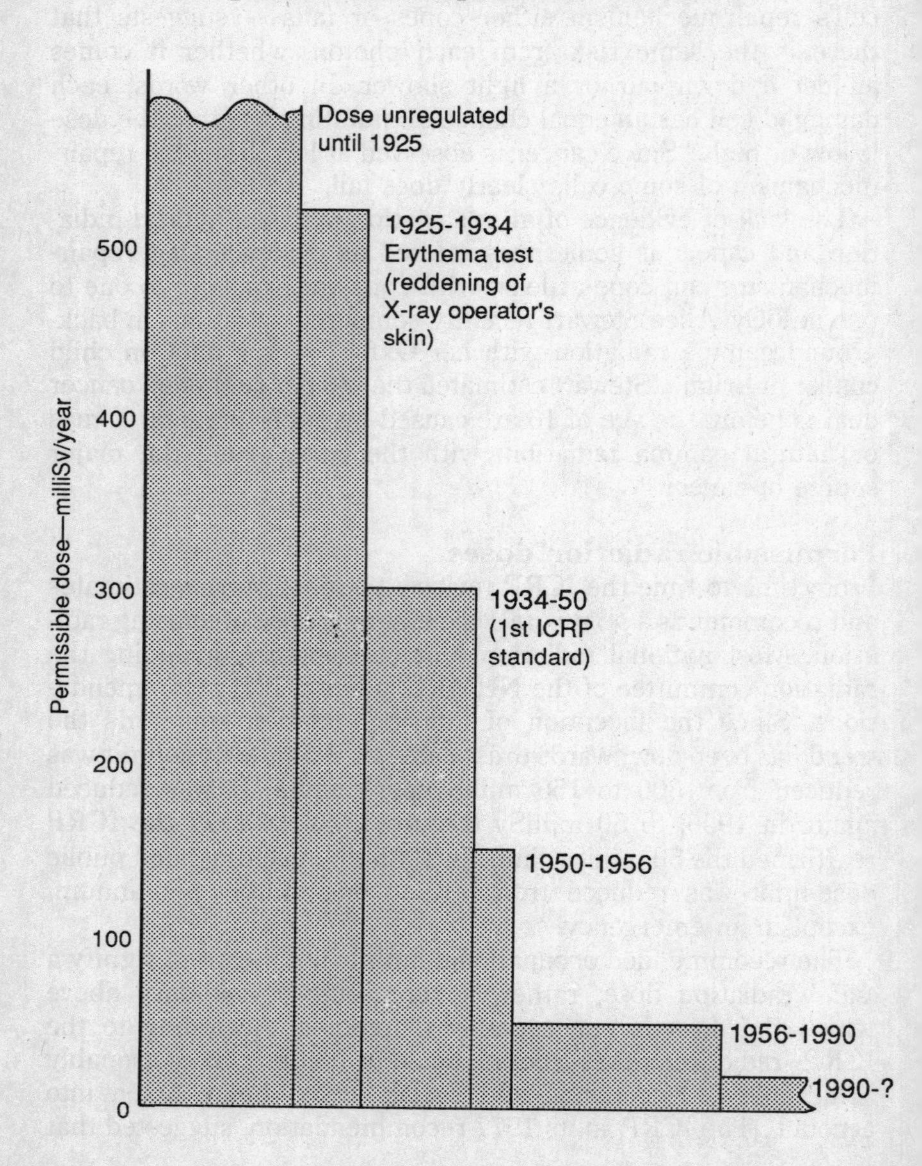

an 'acceptable' average annual occupational dose for workers' health and safety was five milliSv, and that the nuclear industry should be able to operate satisfactorily within this rate of exposure. According to the ICRP's 1977 risk-estimates (see Table 13.3), this rate of exposure would cause about one fatal cancer in about 20 000 workers — although the ICRP identified an occupation involving radiation work as 'safe' where the maximum cancer mortality rate among the workers was two in 20 000. This can be compared with the occupational risks in a range of industries shown in Table 13.2.

This simple comparative approach ignores a difference between radiation workers and those in other industries. Annual radiation doses are additive; to be equitable a worker's *cumulative* radiation dose over his or her lifetime should be used as the basis for estimating the occupational risk. Radiation workers share with workers exposed to toxic chemical agents an added risk, over and above the risks associated with the usual occupational hazards. Most importantly, it is the 50 milliSv level that is legally enforcible in the workplace — not the 'acceptable' 5 milliSv. In fact, many radiation workers are presently exposed to more than 5 milliSv. In Britain over 2 000 workers, half of them in the

Table 13.3: Comparison of cancer fatalities at various dose-levels of an ionising radiation, according to ICRP and BEIR risk-estimates

Annual dose milliSv	**Risk-estimate** Cancer deaths per 100 000 workers		
	ICRP 1977	**ICRP** 1990	**BEIR** 1990
1	1	5	8
5	6	25	40
10	12	50	80
20	24	100	160
50	60	250	400

Until 1990 the criterion used by the ICRP for the nuclear industry to be classed as a 'safe' industry was a *maximum* fatality rate of 10 per 100 000.

nuclear industry, are exposed to more than 15 milliSv per annum. At the Roxby uranium mine in South Australia, workers average 6 milliSv per annum; but 70 per cent of full-time underground workers receive over 10 and up to 30 milliSv. On present evidence many radiation workers and uranium miners are exposed to a high cancer risk (see Table 13.3).

The use of an *average* occupational dose hides the extent of the exposure of many radiation workers. The major part of the collective dose is received by a minority of radiation workers. In Britain seven per cent of nuclear-power workers receive almost half of the collective dose; at the Sellafield reprocessing plant 42 per cent receive 70 per cent of the collective dose.[11] In uranium mining it is the minority of miners at the workface who receive the high doses. The use of averages simply hides the fact. However, spreading the collective radiation dose over a workforce by rotating individuals in the high-risk jobs is no answer to the cancer problem: while the risk for some individuals is reduced, the probable number of cancer fatalities in the workforce remains the same.

In December 1989, the biological effects committee (BEIR) of the US National Academy of Science revised its risk-estimates on the basis of revisions made in cancer mortality rates among the Hibakusha in the ongoing Life Span Study. The revised estimate (BEIR V) was that, for a population of 100 000, exposed to ten milliSv annually, there would be 80 cancer deaths — about four times its previous estimate in 1980. This estimate is about seven times greater than the one used by the ICRP when it recommended permissible exposures in 1977 (see Table 13.1).[12] In February 1990, the ICRP issued its recommendations in the form of a draft, rather than in its usual non-negotiable final recommendation; no doubt the ICRP had become sensitive to criticism of its behind-closed-doors decision-making. In the draft the 1977 estimates of cancer fatality were acknowledged to be four to five times too low.[13] From this it could have been expected that the ICRP would proceed to recommend a proportionate reduction in exposure limits — to, say, ten milliSv. Instead, the draft maintained the 50 milliSv annual dose-limit; although as a concession the ICRP set a subsidiary 100 milliSv limit for any five-year period (that is, 20 milliSv per annum). The public dose-limit was recommended to stay at one milliSv per annum.

Higher risk-estimates raise the question of what the ICRP's 'acceptable' exposure of five milliSv should become in order for the nuclear industry to be judged a 'safe' industry (less than one fatality in 10 000). Simple arithmetic shows that the five milliSv 'acceptable' dose should be reduced to one milliSv (see Table 13.3) — clearly an uneconomic, even if technically achievable, target for the nuclear power and uranium mining industries.

Not unexpectedly, the ICRP has abandoned its 'acceptable' average occupational exposure; instead, it has adopted a concept of *loss of life expectancy* as a basis for comparing work-safety in the nuclear and other industries. The number of days-of-life lost from a fatal radiation-induced cancer is compared with the number of days-of-life lost as a result of a fatal accident. The records of fatal accidents in industry generally show them to occur on average around age forty, whereas cancer deaths occur mostly around age sixty. Accordingly, the ICRP has now taken into account, in its 1990 recommendation, that the loss-of-life expectancy from radiation-induced cancers is less than the usual industrial accident.

What the ICRP has taken into account most of all, though not saying as much, has been the very difficult management and economic problems that would confront industries employing radiation workers should radiation dose levels be lowered to the extent justified by the new data. The other side of that coin, of course, is the health and safety of workers in industrial radiography, in practices involved with medical diagnoses and therapies, as well as those in nuclear industries. Patrick Green, radiation consultant to Friends of the Earth in Britain, has criticised the industry bias of the ICRP's recent recommendations, saying that 'the commission has put the needs of the [nuclear] industry before the need to protect the health of radiation workers and their children'.[14]

The tardiness of the ICRP in recommending lower dose-limits led the National Radiological Protection Board (NRPB), in 1987, to take the initiative and reduce the permissible limit from 50 to 15 milliSv per annum. Sweden has followed Britain's lead, while

Germany has adopted a *cumulative* lifetime limit of 350 milliSv over a 35-year working life; that is, a yearly average of no more than ten milliSv. A number of countries are also moving to reduce the public dose-limit to below the ICRP 1990 draft recommendation of one milliSv. In the United States the environmental protection agency has recommended a permissible public dose of 0.25 milliSv; in Germany the dose has been reduced to 0.3 milliSv. Meanwhile, the Australian occupational dose-limit remains at 50 milliSv and the public dose-limit one milliSv.

A declaration of more than 800 scientists from around the world has called for an immediate five-fold reduction in the annual occupational dose-limit and a cumulative lifetime dose of no more than 100 milliSv. The declaration also proposed the principle that radiation doses should be 'as low as *technically* achievable' instead of the existing 'as low as *reasonably* achievable'.[15]

The ICRP recommendations did not fully consider the report of the British medical research council (Gardner report). That report identified a high incidence of leukaemia among children whose fathers are exposed to radiation at the Sellafield reprocessing plant. The finding has changed the world view of the risks of radiation exposure; it reinforced the argument that the dose-levels in the 1990 ICRP recommendations are much too high.[16] A child of a father exposed to a lifetime dose in excess of 100 milliSv — just five years at ICRP's proposed annual dose-limit — has a one-in-300 chance of developing leukaemia. This finding highlights the way genetic risks for offspring from parental exposures have been underestimated in the setting of radiation standards. The ICRP has said that it will be reconsidering its recommendation in the light of the Gardner report.

IN Australia, a radiation protection committee, appointed by the NHMRC, examines any new risk-estimates and recommendations issued by the ICRP, and then recommends any revisions it believes necessary to existing standards. The revisions are submitted to the federal Department of Health, which publishes them. Federal and state health departments then confer, and eventually decide on permissible dose-levels to be incorporated into various state radiation safety regulations and the federal code of radiation practice for uranium mining and the handling of radioactive wastes.

In the past the NHMRC appears to have done little more than rubber-stamp ICRP recommendations. The NHMRC's response to the 1990 ICRP draft recommendations was, once again, to adopt a position 'similar to that of the international commission on radiological protection. For the moment no revision of the exposure limits is recommended, but users are reminded of their responsibility to ensure that exposures are kept low, especially in those workplaces where significant exposures take place'.

The National Occupational Health and Safety Commission (NOHSC) was established in 1984 'to make Australian workplaces safe and healthy'. The commission is a tripartite organisation composed of representatives from the ACTU, the Confederation of Australian Industry, and federal and state governments. A standards development standing committee of NOHSC has the task of determining priorities for setting occupational safety standards. Asbestos has been given high priority. Many scientists believe that low-level ionising radiation is a potential killer of the same order as asbestos. However, the ACTU appears to have made no real effort to have ionising radiation given any special attention by the standards committee of NOHSC. Instead, the self-elective NHMRC seems to have been left to decide that ICRP recommendations will form the basis of any revised standards. That will mean radiation workers, particularly those in the uranium and mineral-sands industries, will be exposed to radiation levels much higher than have been permitted in Britain and Europe since 1987. The ACTU's apparent indifference to the excessive exposure of workers to ionising radiations contrasts with its previous strong intervention on behalf of workers exposed to RF radiation (see chapter 4).

Risk is relative

Risk can be expressed as the *absolute* risk; that is, the number of *additional* cases of disease for a certain population size. Another way is to estimate the *relative* risk. By this method the risk of additional cases of a disease is expressed as a percentage of the *ordinary* occurrence of the disease in the absence of radiation exposure. A study of uranium miners illustrates how the relative, rather than the absolute, method is much more realistic an indicator when applying a radiation risk derived from one group to another exposed group.[17]

Table 13.4: Lung cancer deaths among smokers and non-smokers in uranium mines

	Lung cancer deaths per 100 000 per year
Non-smoking miners	12.5
Non-smokers, uranium miners	20.0
Heavy smokers, non-uranium miners	265.0
Heavy smokers, uranium miners	700.0

In Table 13.4 we see that the rate of lung cancer among non-smokers is increased by mining from 12.5 to 20, or an increase of 7.5 cancers per 100 000 per year. Taking the increase of 7.5, which can be attributed to mining uranium, and adding it to the non-miner heavy-smoker death rate of 265 per 100 000 per year, we get a rate of 265 + 7.5 cancers. Thus, by the absolute risk method, the mortality rate among smoking miners should be 272.5. But the actual number of lung cancer deaths among this group was found in the study to be 700. This makes the excess rate for heavy smokers 435, not 7.5, cancers per 100 000.

Now, if instead of doing our calculation by the absolute risk method, we determine the relative risk in the first group, we get an entirely different assessment. The relative risk of mining for non-smokers is 7.5 divided by 12.5 — or 60 per cent. Applying this percentage to the heavy-smoker group, we find that mining should increase their cancer death rate by 159 — that is, to 424 per 100 000 each year. This still underestimates the cancer risk of the smokers; nonetheless, it is a much more realistic indication of risk for us to go by.

The gap between the calculated and actual number of lung cancers reflects a problem inherent in estimating the lifetime risk. Some people living at the time of the survey may still die of lung cancer. It could also be explained by a *synergistic* interaction between smoking and mining. When this happens the health hazard of the two factors acting together is greater than the sum of the hazards of each factor acting separately.

Radiation and heredity

Children inherit mutational damage to the family line, although it may show up only in later life. Just as a cancer may lie dormant for thirty or more years after a child has been irradiated, before its clinical diagnosis, so too an inherited predisposition to a disease may not show up until later in life.

There is difficulty in distinguishing between genetic damage caused by artificial radiation and that which occurs because of other environmental factors or spontaneously from natural causes. The only direct evidence available to radiation advisory committees, such as the ICRP and BEIR, for predicting transmittable genetic damage which is induced by radiation, comes from studies on other mammalian species — mostly mice. The problem is made difficult by there being so little known about how large is the genetic component in the overall incidence of common diseases. BEIR has estimated the incidence of all types of genetic disorder to be 107 000 for every million babies born. Of these, 17 000 fall into the category of *severe* genetic disorders. The remaining 90 000 are the *constitutional* and *degenerative* diseases (the so-called irregularly inherited disorders) such as anaemia, diabetes, schizophrenia and epilepsy (see Table 13.5).

Table 13.5: Estimate of the incidence of constitutional and degenerative diseases

Disease	Incidence per million live-born individuals
Dominant and X-linked	10 000
Recessive	1 100
Chromosomal	6 000
Irregularly inherited	90 000
Total	107 100

Source: BEIR III, 1980.

Heart disease, ulcer and cancer are not included, even though these diseases are known to have a significant genetic component. It is difficult to understand why the BEIR committee has been so selective. The majority of estimates of the genetic component in known diseases has been between 25 and 50 per cent.[18]

The ICRP says its permissible exposures are set so that there is a negligible probability of *severe* genetic disorder. According to ICRP, a permissible genetic dose is that dose which, if received yearly by each person from conception to average childbearing age (taken as 30 years), would result in an *acceptable* burden to the whole population of genetic disorders. 'That might be paraphrased', Bertell said, 'to say that the general public (government) may be willing to accept the number of blind, deaf, congenitally deformed, mentally retarded and severely diseased children resulting from the permissible exposure level'.[19] In their wisdom, Bertell said, expert committees have judged the value of medical radiation, nuclear power and nuclear weapons as worth society taking these genetic risks. What Bertell had to say has been given greater poignancy by the recent findings of the Gardner research team: fathers who are exposed within permissible radiation levels at the Sellafield nuclear plant in Britain have a high probability of bearing children who will suffer from leukaemia.

Workers and the community are told that the chances of severe genetic damage from radiation exposure from nuclear and medical activities is slight; but little is said about premature deaths, including many among the not-so-old, from an inherited predisposition to the constitutional and degenerative diseases. Experts put the focus of the health costs of radiation exposure on cancers and severe genetic diseases. Disputes among radiation experts about radiation hazards have mostly to do with the numbers of fatalities and the more apparent diseases. But there is other suffering and anguish in families that has to do with cardio-vascular diseases; diabetes; loss of immune resistance — especially among the very young and old — to contagious disease; acceleration of cancer; and other mild but debilitating health effects that diminish the quality of life. The contribution made by radiation emissions from nuclear activities and overuse of medical radiations to these so-called irregularly inherited disorders is unknown — partly because of the large scale of the epidemiological studies needed to follow the 'irregular' patterns of inheritance, but also because the advisory bodies recommending on radiation exposures show little enthusiasm for such studies.

Pioneer geneticist Hermann Muller warned of the prospect of a gradual reduction of the survival ability of the human species as successive generations are exposed to ionising radiation. Therein lies a moral challenge to stop nuclear activities and restrain medical exposures to the barest minimum.

NON-IONISING RADIATIONS

At first encounter, the controversy over the health effects of non-ionising electromagnetic radiation (EMR) can be more perplexing than the controversy over ionising radiation. With the more energetic ionising radiations, scientists are generally agreed on the nature of the hazards, or at least the severe ones of cancer and genetic disorder. Disagreements occur largely over the extent of the risk. The experts mostly talk the same jargon.

Controversy over non-ionising EMR is much more confused, reflecting less coherent research aims. It is only in recent years that this research field has grown into a multi-disciplinary science of bioelectromagnetics. Research ranges widely over cell biology, the epidemiology of cancer, and behavioural studies of animals and humans. The diverse phenomena investigated include metabolic disturbances in the nervous and endocrinal systems, psychological stress, changes in the pineal organ relating to bird navigation, calcium efflux in brain cells, and promotion of cancer-cell growth. All this presents a somewhat mixed bag of scientific topics and jargon.

Many of the health effects attributed to non-ionising EMR are not quantifiable; they cannot be reckoned in body counts. They are more appropriately described holistically as a certain loss of good health and wellbeing. Such health detriments have not been taken seriously by expert committees that decide on safety. At its roots, the controversy is as much a matter of philosophy as it is of science.

Windows and waveforms

Two new hypotheses of bioelectromagnetism were described during the 1970s. Both have aroused debate in the scientific community. One is that non-ionising EMR interacts with living tissue in narrow frequency bands to bring about certain athermal bioeffects. Biological activities which occur only at certain intensities have also been observed; the activity falls away with either an increase or a decrease of intensity. These specific frequencies and intensities are referred to as *windows*.

The concept of EMR windows developed out of the observation that the flow of calcium in brain cells is altered by radiation from a narrow band centred on 16 Hz. Other windows have since been identified with changed rates of cytolysis (the killing of cells) at 50 Hz, and with changes in neurotransmitters in the nervous system at 500 Hz. There are natural analogues of windowed effects of EMR. The visible-light band provides a window for photosynthesis and vegetative growth. Through the same window the eye is enabled to interpret weakly reflecting light to create coloured images in perspective. Some micro-organisms have a window on natural ultraviolet light.

The other concept is that *waveforms* act as carriers of signals that convey information (or misinformation) to living cells to trigger athermal biological activity. Waveforms can be continuous, modulated or pulsed. Athermal bioeffects appear to occur only with ELF waves or with ELF as the modulating or pulsing frequency of carrier waves.

The information content of the waveform and the special significance of pulsing and modulation have been demonstrated by behavioural tests. Animal testing indicates that the activating frequency can be specific for species behaviour. In humans, the most biologically active frequencies appear to be in the same frequency range as the body's natural biosignals.

The concept of biological activity dependent on frequency and intensity windows, and on waveform as an information source, presents difficulties for assessing radiation health-risks. The effect of chemical toxic agents is generally proportional to the dose: the body is able to recover below a certain dose. According to the window concept, non-ionising radiation can be harmful only at a specific intensity and frequency. This rules out a simple dose-response relationship of the kind that applies for higher energy radiation.

Bioeffects v health hazards

Exposure to certain levels of non-ionising EMR, especially microwaves, has the potential to overheat the body. The thermalist view, which prevailed for a long time, is that heat poses the only health hazard. The absorbed heat is likened to a fever, something with which the body can cope except in extreme cases. By the 1980s the thermalist hypothesis faced a strong challenge from the accumulating evidence of athermal bioeffects.

With the existence of athermal bioeffects now beyond question, thermalists nowadays argue that they cause only *transient* disturbances to body functions. The disturbances, they say, are probably no greater than fluctuations in the body's physiological and biochemical processes commonly experienced from all kinds of external stimuli. The Australian standard on RF health and safety says, 'It has been demonstrated that low-level, long-term exposure can induce a variety of effects in the nervous, haematopoietic and immune systems of small animals. Such exposure may influence the susceptibility of animals to other influencing factors. Thermal mechanisms seem inadequate to account for these and other effects'. Thus, the position of thermalists remains essentially unaltered: the only health risk is irreversible damage from severe overheating of the body.

Athermalists acknowledge that the occurrence of bioeffects does not in itself establish a health hazard. However, since some athermal bioeffects occur in the body's most sensitive organs — the endocrinal, immune and nervous systems — the probability is that they do affect health, at least in the sense of well-being.

There are other indicators of a relationship between bioeffects and ill health. Changes in the rate of hormone secretions, which have been linked to factors which promote cancer, can be brought about by ELF radiation; they could, therefore, explain the evidence that such radiation promotes malignant growth. Non-ionising EMR is being used to heal intractable bone fractures; evidently, it brings about fundamental changes in the regeneration processes in bone cells.

Many of the demonstrated bioeffects may be transient in experiments. But in real life they can be protracted, because people are exposed to the radiations for long periods — often, for most of their lifetime. Behavioural experiments are now showing that impairment of learning ability and memory in animals may be more persistent than once thought.

The short duration of some of the observed bioeffects is indicative of the body's ability to handle the stress to which its physiological and neurological processes have been subjected. But chronic stress from protracted EMR exposure may well tax the adaptive capacities of the body. Also, little consideration is being given to the additive effect of the increasing number of environmental stressors — radiations or chemicals — or the synergism that is likely to come into play between some of these stressors.

Epilogue

Democratising the setting of standards

RADIATION is the most studied of the factors known to affect environmental health. Yet uncertainty remains, and will continue to remain, about the nature and extent of the health hazards it represents. After weighing the technical information, community perceptions of what is an acceptable risk must rely a good deal on people's own value judgements. In a well-informed democracy, communities are not without a sense of responsibility for the needs of others. If people are volunteered sound information then a balance can be struck between what is best in the interests of both community and nation.

For this to occur, the present institutional processes need to be opened up. The process whereby a local community responds to an environmental impact statement has proved quite inequitable in any conflict involving a project proponent and a community. The real decisions have almost invariably been made either in closed government committees or after private consultations between administrators, corporate officials and government-appointed technical experts.

Legislative steps have been taken federally and in the states to involve worker representatives in the setting and implementation of occupational radiation and other health and safety standards. It is no more than equitable that communities, which now live in an environment so much affected by industrial pollution, should have some say in the setting and enforcement of environmental health standards. Of course, while worker participation in occupational health and safety has a long history of struggle, community participation in such matters has only recently become a live issue. However, the demand is growing for affected communities to be represented in the decision-making processes which concern any significant alteration to their environment.

It will certainly prove difficult to decide who ought to represent the concerns of the wider community at a national level; in the workplace, where trade union representation is already in place, the matter is simplified. However, at a community level, health centres and health councils (at least in Victoria) already function with joint committees of government and community representatives, and they will undoubtedly be playing a greater part in environmental health. Community action groups keep a watching brief on health and environmental matters; and health officers in local government are involved in environmental health. The basis exists for a national network out of which could come community representation in the regulatory processes of environmental health at state and federal levels.

It is not just a matter of community participation on the basis of ready-made safety standards. It is at least as important for lay representatives to be able to raise questions about environmental pollution which they see as needing investigation. This means more face-to-face interaction between lay and scientific communities — not only in enforcing existing health and safety standards, but also in directing the research to provide a relevant data base.

Presently, much research is devoted to determining the biological activity of radiation absorbed by the human body — as if it is an inevitable phenomenon and not artificial. The health consequences are being researched, while the ways and means of lessening the radiation pressure on the environment are being neglected.

We need to explore social alternatives that avoid potential health-damaging environmental pollution, such as hazardous radiations. Obvious first steps should include support for solar, rather than nuclear, sources of energy; and the selection of communications systems that minimise electromagnetic radiation exposures — such as optical fibres.

The environmental impacts of non-nuclear radiation sources can be reduced if the political will is there to suitably site and shield the installations. We can do without the radiations emanating from the activities of the nuclear power and weapons industries. Above all, we need to question the economic myths that motivate our growth-oriented and ever more polluting society. Only then will we seriously begin to re-establish a genuinely healthy relationship with our natural environment.

Appendix

Where radiations come from

RADIATIONS come from atoms. To understand how, we need to know a little about the structure of atoms. The ancient Greek philosopher Democritus was the first to speculate that all things in the universe were made up of invisibly small particles. He called them 'atoms' — the Greek word for indivisible. In the seventeenth century Isaac Newton, the founder of modern physics, still described atoms as being 'so hard as never to wear or break into pieces'.

The idea of the solid indestructible atom did not survive for very long after 1896 when Henri Becquerel discovered that uranium emits radiation unceasingly. In 1897 English physicist J J Thomson described cathode rays as a stream of negatively charged particles which he called *electrons*. They must be stripped, he speculated, from the metal atoms of the negative electrode in the cathode-ray tube. Thomson thought of electrons embedded in atoms like currants in a bun.

It was the discovery of radium and its more powerful rays by Marie and Pierre Curie in 1898 that really cast doubt on the supposed inertness and solidity of the atom. Specks of radium, a million times more active than uranium, emitted a startling quantity of heat and radiation. Dramatic events were certainly occurring *inside* radioactive atoms.

New Zealander Ernest Rutherford, once an assistant to Thomson and a founder of nuclear physics, began to use radium's alpha particles to probe inside atoms. When Rutherford fired alpha particles at gold foil most of the particles passed right through, suggesting mostly empty space. Some particles were deflected at varying angles. What really excited Rutherford was the occasional particle that bounced straight back. 'It was quite the most incredible event that ever happened to me in my life', Rutherford wrote. 'It was almost as incredible as if you had fired a 15-inch shell at a piece of tissue paper and it came back and hit you'.

The picture of atomic structure to emerge from this and other experiments was of a planetary system of electrons orbiting a nucleus. It had been direct hits on gold nuclei that had caused alpha particles to bounce straight back. Rutherford's experiment amounted to using radium nuclei as a source of alpha particles to fire at the nuclei of gold atoms as the target.

The planetary model was at first only a hunch; but it became the rudimentary base from which to start thinking about atoms. From the model we can come to some understanding of the forces behind radiation emissions. The nucleus is made up of neutrons and protons. The orbiting electrons have a negative charge. Since atoms in their normal state are electrically neutral, the protons must have positive charges to balance the electrons' negative charges. The number of orbiting electrons equals the number of protons in the nucleus. The neutrons, as their name implies, are neutral. But as we shall see they do much more than add ballast to atomic nuclei.

Inside the atom, Rutherford said, electrons are 'like a few flies in a cathedral'. The diameter of an atom is about ten thousand times the diameter of its nucleus. The mass of an electron is only a two-thousandth of that of either a proton or a neutron. So virtually all the mass of an atom is packed into its relatively tiny, dense nucleus.

While the planetary model explained many properties of atoms, especially the way they behaved chemically, it also posed problems about why they remained as stable as they did. According to physical laws at the time, an electron could move in a closed orbit only if it radiated energy all the time. But this means a continuous loss of energy, so that it must lose momentum and fall into the positive nucleus. Clearly this did not happen.

Danish physicist, Niels Bohr, who worked with Rutherford for a short time in 1911, came up with an answer. He studied hydrogen, the simplest atom, with a single proton in the nucleus and a single electron in orbit. He applied Einstein's concept of 'packets' or *quanta* of energy (see box: Radiation energy is delivered in 'packets', in chapter 13). After two years of theorising, Bohr put forward a model of the hydrogen atom in which the electron could only circle in one of a number of fixed orbits, which did not require it to radiate energy.

The electron normally stays in the nearest orbit to the nucleus allowed by its energy. If the atom is excited by incoming energy the electron jumps to an orbit further out. However, the electron must absorb a certain energy quantum to occupy a particular orbit. How far out it jumps depends on the size of the quantum. As the excited atom loses energy the electron returns to the orbit appropriate to its 'ground' state. With each jump to a lower orbit, electromagnetic radiation is emitted. The frequency of the emitted radiation is fixed by the size of the jumps. The quantum of energy emitted equals the quantum absorbed.

Bohr chose the orbits according to the frequencies measured for radiation emitted by hydrogen. He did not explain why they remain fixed. French scientist Louis de Broglie has given a possible explanation as to why electron orbits exist only at certain radii. If an electron was associated with standing waves (like on a guitar string) joined end to end, then for any whole number of waves only certain orbits would fit for any particular frequency. It is like the way a snake head-to-tail fits only a certain circle. Now we have something quite different from a picture of particles quietly circling a nucleus. As electrons orbit at an amazing speed of 600 kilometres a second, atoms take on the appearance of clouds of concentrated energy. The great speed in so confined a space makes an atom look solid just as a whirring propeller appears to be a disc.

The Bohr atom can explain how the frequency of electromagnetic radiation varies. Electrons shifting between the outermost orbits produce energy with frequencies in the visible light and infrared bands. Shifts closer to the nucleus produce higher frequency ultraviolet light. The energy for these electron shifts can come by heating, such as in an incandescent electric light globe. X-rays are produced from shifts of the innermost electrons in heavy atoms, such as those of metals, which have absorbed energy from the high-speed electrons generated in a cathode-ray tube.

The radio frequencies which lie below the infrared band on the spectrum are produced when electrons oscillate in an electric spark or a circuit like an antenna. Studies on free-flowing electrons in circuits suggest that they behave as matter/waves or 'wave packets' as they do in atomic orbits, losing energy when they give out radiation.

So much for the electromagnetic radiations emitted by electrons. What of those radiations from radioactive substances, which first set physicists on the path of atomic discovery?

The forces binding the protons together in the nucleus cannot be electrical because the positive charges on protons are mutually repelling. Neither can electrical forces be invoked to explain the binding of neutrons because they do not have electric charges. An explanation is found in the way a free proton fired at a nucleus can be absorbed into it despite both carrying positive charges. As the proton comes within about one million millionths of a millimetre of the nucleus it is overpowered by an extremely short-range force that binds it into the nucleus. The nuclear binding force is greater than any other force known in nature — a kind of atomic 'super-glue'.

Nonetheless, neutrons and protons do sometimes emigrate from the nuclei of radioactive atoms and in the process give off energy. Albert Einstein provided an explanation for the liberation of the energy in 1905 while still a patents office clerk. The energy came from matter! He expressed this in his famous equation $E=mc^2$.* As a radioactive atom decays, a little of the mass of the nucleus is converted to energy. The quantity of energy corresponds to the mass of the lost matter multiplied by the astronomically large number of the speed of light squared (multiplied by itself). This means that the slightest particle of matter holds an enormous store of energy. Radioactive atoms release their stored energy as alpha, beta and gamma radiation and the occasional neutron.

* 'E' is the energy in joules; 'm' is the mass in kgms; 'c' is the speed of light, which is 300 million metres per second. Translated into more familiar terms, the energy from a gram of matter would keep a one-kilowatt radiator running for nearly 3 000 years.

Familiar analogies are helpful in illustrating the spatial arrangements and transformations of electrons. But once we look *inside* the nucleus we must somehow move beyond the images we create in our 'real' world. We cannot fully understand the reality of nuclear matter. We can only resort to metaphor in our thinking about nuclear phenomena.

There is constant interaction between protons and neutrons inside a nucleus. This interaction somehow involves incessant interchange of mass and energy through an intermediary particle, called a *meson*. It is this mediation of the meson which brings about a levelling of protons and neutrons so that they can be thought of as *nucleons*. The meson is in fact only one of some 200 particles now known to be constantly coming and going inside the nucleus. It is a melee which defies description. Not surprisingly, different models have been proposed to represent nuclear phenomena.

One model was proposed by Neils Bohr in the 1930s. It was the 'liquid-droplet model', in which nucleons move about in much the same way as molecules do in a liquid drop. If an ordinary liquid droplet is heated the molecules become agitated and begin breaking away from each other. They may boil off gently or the droplet may divide, depending on the nature of the substance and the quantity of heat energy absorbed. A 'droplet' nucleus can be pictured as behaving similarly, and is extraordinarily dense. If enough nuclei could be taken out of their spacious atoms and packed like oranges into a box, a cubic millimetre of this pack (the volume of a pin's head) would weigh 200 000 tonnes. The nucleons respond to their dense packing by moving at an estimated 70 000 kilometres a second.

Because the nuclear binding force has such a very short range, the more vigorously vibrating particles in the less stable nuclei have some chance to pry themselves loose — despite the powerful binding energy. The nucleus of a radionuclide has this kind of instability. A pair of protons and a pair of neutrons may team up and gather the momentum needed to push ever so little beyond the reach of the forces. Once that happens, the repulsion between the positive charges gets the upper hand and an alpha particle carrying two positive charges flies off.

The emission of an electron (beta ray) from a positively

charged nucleus had been surmised for a long time to occur because a neutron was a union of a negative electron and a positive proton. The beta particle would result when a neutron broke into its two parts. It has been demonstrated (using a precision mass spectrometer) that in fact the neutron has a slightly greater mass than a proton. However, the greater mass of the neutron does not come from an electron, because it is virtually massless. It comes in accordance with Einstein's equation for the conversion of energy to mass from the energy of the electron-proton union. When this conversion of energy into mass is reversed the neutron flies apart and becomes a proton plus an electron in a form of a beta ray.

A neutron does not have the advantage of an electrical charge to shove itself away. In fact, spontaneous emission of neutrons is a very much rarer event in radioactive material than emission of alpha particles. But it happens in, for example, the fission of uranium atoms in nuclear reactors and nuclear bombs.

When an alpha or a beta particle is given off, the atom undergoes a *transmutation*: that is, a complete change in its chemical nature to become a different element. As the excited state of the nucleus subsides, excess energy is given off as gamma radiation. The energy of the gamma radiation is characteristic of the radionuclide from which it is emitted: this enables the identification of particular radionuclides in radioactive fallout, such as that which occurred after the Chernobyl accident.

The 'droplet' model can also be used to explain the way a nucleus splits or fissions. When a neutron enters a fissionable nucleus it creates instability, or sets up a wobble. For example, the shape of the nucleus can fluctuate between the shape of a sphere and the shape of a dumbell. The two halves of the dumbell might then separate a little too far and leave each other for good. This splitting (fission) is similar to what we see happen when a floating liquid drop is set wobbling.

Bibliography

Referenced literature and a selection of background reading used in the preparation of the book.

Books

Ball, Howard, *Justice Downwind*, Oxford University Press, 1986.

Becker, Robert & Marino, Andrew, *Electromagnetism and Life*, State University Press of New York, 1982.

Becker, Robert O. & Seldon, Gary, *The Body Electric*, William Morrow & Co., 1985.

Bertell, Rosalie, *No Immediate Danger — Prognosis for a Radioactive Earth*, The Women's Press, London, 1985.

Blakeway, Dennis & Lloyd-Roberts, Sue, *Fields of Thunder — Testing Britain's Bomb*, Unwin Paperbacks, 1985.

Brodeur, Paul, *Zapping of America — Microwaves, Their Deadly Risk and Cover-up*, W.W. Norton, New York, 1977.

Caufield, Catherine, *Multiple Exposures — Chronicles of the Radiation Age*, Secker & Warburg, London, 1989.

Committee for the Compilation of Materials on Damage by the Atomic Bombings on Hiroshima and Nagasaki, *Hiroshima and Nagasaki — The Physical, Medical and Social Effects of the Atomic Bombings*, Hutchinson, London, 1979.

Gofman, John W., *Radiation and Human Health*, Sierra Club Books, San Francisco, 1981.

Gofman, John W. & O'Connor, Egan, *X-Rays — Health Effects of Common Exams*, Sierra Club Books, San Francisco, 1986.

Gofman, John W., *Radiation-Induced Cancer from Low-Dose Exposure: An Independent Analysis*, 1990. Available from Committee for Nuclear Responsibility, P.O. Box 11207, San Francisco, California 94101, USA.

Gould, Jay M. & Goldman, Benjamin A., *Deadly Deceit*, Four Walls Eight Windows, New York, 1990.

Hersey, John, *Hiroshima*, Penguin, 1946.
Kelen, Stephen, *I Remember Hiroshima*, Hale and Iremonger, Sydney, 1983.
Lapp R.E., *The Voyage of the Lucky Dragon*, Penguin, 1958.
Mabusa, Masuji, *Black Rain*, John Martin, London, 1969.
Marino, Andrew & Ray, Joel, *The Electric Wilderness*, San Francisco Press, 1986.
McSorley, Jean, *Living in the Shadow. The story of the people of Sellafield*, Pan Books, London, 1990.
Medvedev, Zhores A., *The Legacy of Chernobyl*, Basil Blackwell, Oxford, 1990.
Milliken, Robert, *No Conceivable Injury. The story of Britain and Australia's atomic cover-up*, Penguin, 1986.
Pringle, Peter & Spigelman, James, *The Nuclear Barons*, Michael Joseph, London, 1982.
Prins, Gwyn (Ed.), *Defended To Death*, Penguin, 1983.
Rhodes, Richard, *The Making of the Atomic Bomb*, Simon and Schuster, London, 1986.
Rotblat, Joseph, *Nuclear Radiation in Warfare*, (Stockholm International Peace Research Institute), Taylor & Francis, London, 1981.
Russell Jones R. & Southwood R. (Eds), *Radiation and Health — The biological effects of low-level exposure to ionizing radiation*, Wiley, Brisbane, 1987.
Schubert, Jack. & Lapp, Ralph E., *Radiation — What It Is and How It Affects You*, Heinemann, Melbourne, 1957.
Smith, Joan, *Clouds of Deceit — The Deadly Legacy of Britain's Bomb Tests*, Faber and Faber, 1985.
Steneck, Nicholas, *The Microwave Debate*, MIT Press, Cambridge, Mass., 1984.
Steneck, Nicholas (Ed.), *Risk/Benefit Analysis — The Microwave Case*, San Francisco Press, 1982.
Tame, Adrian & Robotham, F.P.J. (Rob), *Maralinga*, Fontana, Melbourne, 1982.
Townsend, Peter, *The Postman of Nagasaki*, Penguin, 1984.
Wasserman, Harvey & Solomon, Norman, *Killing Our Own*, Delacorte Press, 1982.
Wilkes J., *Field of Thunder*, Friends of the Earth, Melbourne, 1981.

Reports and reviews

Adey W.R., 'Tissue Interactions with Non-ionising Electromagnetic Fields', *Physiological Reviews*, Vol. 61, No. 2, 1981.

Adey W.R., 'Nerve Cell Membrane and Intracellular Communication in Brain Tissue', *Fourth Abbie Memorial Lecture*, University of Adelaide, 1985.

American Institute of Biological Sciences, *Biological and Human Health Effects of Extremely Low Frequency Electromagnetic Fields. Post 1977. Literature Review*, Report of the Committee on Biological Effects of Extremely Low Frequency Electromagnetic Fields, Arlington, Va, USA, 1985.

Armstrong B., *A Review of Epidemiological Data on the Effects of Extremely-Low-Frequency Electromagnetic Fields (ELF) on Human Health*, Department of Medicine, University of Western Australia, December, 1986.

Badash L., 'Radium, radioactivity and the popularity of scientific discovery,' *Proceedings of the American Philosophical Society*, Vol. 122, pp. 145-154, 1978.

Bertell R., *Handbook for Estimating Health Effects from Exposure to Ionising Radiation*, Institute of Concern for Public Health, Toronto, Ontario, Canada, 1984.

Brodeur P., 'Annals of Radiation — The Hazards of Electromagnetic Fields,' *The New Yorker*, 1989. Part I: 'Power Lines,' pp. 51-88, 12 June; Part II: 'Something is Happening,' pp. 47-73, 19 June; Part III 'Video Display Terminals,' pp. 39-68, 26 June.

Cooper M.B., Ralph B.J. & Wilkes M.J., *Natural Radioactivity in Bottled Mineral Water available in Australia*, Australian Radiation Laboratory, Report TR 036, 1981.

Delpizzo V. & Keam D.W. (Ed.), *Carcinogenic Potential of Extremely Low Frequency Magnetic Fields*, Workshop Proceedings, Australian Radiation Laboratories, 1988.

Delpizzo V., 'An Evaluation of the Existing Evidence on the Carcinogenic Potential of Extremely Low Frequency Magnetic Fields,' *Australasian Physical & Engineering Sciences in Medicine*, Vol. 12, No. 2, pp. 55-68, 1989.

Graves H.B., *A Review of the State Electricity Commission of Victoria (SECV) Transmission Line and Station Design Practices in Relationship to the Health and Safety of People*. Prepared for Minister for Health, Victorian Government, Aust., Mar., 1986.

Green P., *International Commission on Radiological Protection*, Greenpeace, London, 1985.

Hamersley H., 'Radiation Science and Australian Medicine 1896-1914,' *History Records of Australian Science*, Australian Academy of Science, 1981.

Hamersley H. & Moroney J., *Hedley R. Marston, FRS and the Atomic Weapons Safety Committee — Controversy over fallout from British Nuclear Tests in Australia in 1956*. Prepared for the Royal Commission into British Nuclear Tests in Australia.

IEEE Power Engineering Society. Service Centre, *Biological Effects of Power Frequency Electric and Magnetic Fields*, 445 Hoes Lane, Piscataway, N.J., USA.

Johnson G., *Collision Course at Kwajalein — Marshall Islanders in the Shadow of the Bomb*, Pacific Concerns Resource Centre, P.O. Box 9295, Newmarket, Aukland, New Zealand.

Joyner K.H. & Bangay M.J., *RF Dielectric Heaters. Survey Results and Guidelines for Limiting Occupational Exposures to Radiofrequency Radiation*, Australian Radiation Laboratory, Report TR052, 1983.

Joyner K.H., *et al.*, *Electromagnetic Emissions from Video Display Terminals (VDTs)*, Australian Radiation Laboratory, Report ARL/TR067, 1984.

Kerr C.B., Bently K., Keam D.W. and Robotham R.P.J., *Report for the Expert Committee on the Review of Data on Atmospheric Fallout Arising from British Tests in Australia*, AGPS, 1984.

McGinty L. & Atherly G., 'Acceptability versus Democracy,' *New Scientist*, pp. 323-5, 1977.

Mathews J., *Health and Safety at Work — Australian Trade Union Safety Representatives Handbook*, Pluto Press, Sydney, 1985.

McMillan I., *Electromagnetic Fields, Electric Power and Public Health*. A community resource document based on the Victorian experience during 1985-7. Collingwood Community Health Centre, Sackville St., Collingwood, Victoria, Australia, 1987.

Morris N.D., *An Examination of the Distribution of Patient Doses from Diagnostic X-ray Procedures*, Australian Radiation Laboratory, Report TR051, 1983.

Muller H., 'Radiology and Heredity,' *American Journal of Public Health*, vol. 54, pp. 42-5, 1964.

New York State Powerline Project Scientific Advisory Panel, New York State Public Service Commission (now the New York Power Authority), *Biological Effects of Powerline Fields*, July, 1987.

Norris N.D., *An Examination of the Distribution of Patient Doses from Diagnostic X-ray Procedures*, Australian Radiation Laboratory, Report TR051, 1983.

Powerline Review Panel, *Final Report to the Victorian Government*, Melbourne, 1989.

Repacholi M.H., 'Hazards of Non-Ionising Radiations,' *Radiation Protection in Australia*, Vol. 2, pp. 3-16, 1984.

Repacholi M.H., 'Medical Ultrasound: Is There a Hazard?' *Australian Physical & Engineering Sciences*, Vol. 6, pp. 64-7, 1983.

Richardson J.F., *Australian Radiation Laboratory — A concise History, 1929-1979*, AGPS, Canberra, 1981.

Rotblat J., 'A tale of two cities — Hiroshima and Nagasaki — has taught us a lot about the medical effects of radiation,' *New Scientist*, 7 January 1988.

Royal Commission into British Nuclear Tests in Australia, *Vols. 1 and 2; Conclusions & Recommendations*, AGPS, Canberra, 1985. The Royal Commission exhibits and transcript are lodged in the Australian Archives, Canberra.

Standards Association of Australia, *Maximum Exposure Levels — Radio-Frequency Radiation — 300kHz to 300GHz*, Australian Standard 2772-1985.

State Electricity Commission of Victoria (SECV). *Environment Effects Statement: Proposed 220kV Transmission Line, Brunswick-Richmond; Part B: Biological and Human Health Effects from Transmission Lines and Stations in Victoria*, 1985.

Steneck N., Cook H.J., Vander A.J. & Kane G.L., 'The Origins of US Safety Standards for Microwave Radiation,' *Science*, Vol. 208, pp. 1230-7, 1980.

Steneck N.H., 'Subjectivity in standards: the case of the ANSI C95.1-1982,' *Microwaves and RF*, pp. 137-9, 141-2, 164-6, May 1983.

United Nations Environment Programme, *Radiation — Doses, Effects, Risks*, P.O. Box 30552, Nairobi, Kenya, 1985.

Urquart J., *Leukemia and Nuclear Power in Britain — The Evidence so Far*, Friends of the Earth, London, UK.

World Health Organisation, *Radiological Examination of Drinking Waters*, EURO Report No. 17, 1979.
World Health Organisation Scientific Group, *A Rational Approach to Radiodiagnostic Investigations*, 1983.
World Health Organisation, *Extremely Low Frequency (ELF) Fields*, Health Criteria No. 35, Geneva, 1984.
World Health Organisation, *Ultrasound*, Environmental Health Criteria 22, 1982.

Notes

PART I

Chapter 1

1. Roentgen W.C., 'On a New Type of Rays', *Alembic Club Reprint* — No. 22, London, 1958.
2. Clendinnen F.J., 'Notes on the Roentgen Rays', *Intercolonial Medical Journal*, p. 497, 20. Aug. 1896.
3. Schubert J. & Lapp R.E., *Radiation — What It Is and How It Affects You*; quoted on p. 14; Heinemann, 1957.
4. Hamersley H., 'Radiation Science and Australian Medicine 1896-1914', *History Records of Australian Science*, Australian Academy of Science, 1981.
5. Bowker R.S., 'Case of Bullet Wound of Right Thigh with Successful Removal of Bullet, with the Assistance of X-Rays of Roentgen', *The Australasian Medical Gazette*, p. 512, 21 Dec. 1896.
6. Taylor L.S., *Radiation Protection Standards*, p. 14, CRC Press, Cleveland, 1971.
7. See *Ref.* 3, p. 16.
8. Thomson J., 'X-Rays in Surgery', *The Australasian Medical Gazette*, p. 531-2, 20 Nov. 1897.
9. March H.C., 'Leukemia in Radiologists in a 20-Year Period', *American Journal of the Medical Sciences*, Vol. 220, p. 282-6, 1950.
10. Warren S., 'Longevity and Causes of Death from Irradiation in Physicians', *Journal American Medical Association*, Vol. 162, p. 464-8, 1956.
11. Macht S. & Lawrence P., 'National Survey of Congenital Malformation Resulting from Exposure to Roentgen Radiation', *American Journal of Roentgenology*, Vol. 72, pp. 442-66, 1955.
12. Gofman J. & O'Connor E., *X-Rays: Health Effects of Common Exams*, p. 4, Sierra Club Books, 1985. The book provides a practical guide to the risk of common medical and dental diagnoses. See also Comparative *Risk-Cost Benefit Study of Alternative Energy Sources of Electrical Energy*, pp. 4-14, Government Printing Office, Washington, 1978, in which the US American Energy Commission estimates that a child is

3 to 6 times more likely than an adult to contract cancer from the same dose of ionising radiation.

13. 'The Use of Irradiating Apparatus and Radionuclides in Veterinary and Government, Institutions', *Fifth Report*, p. 1, Victorian Consultative Council on Radiation Safety, 1982.
14. 'Safety of some X-ray equipment doubted', *Age*, 15 June 1982. See also 'Roper warns of danger of radiation,' *Age*, 11 Aug. 1982.
15. Gabrielle C.L., A letter to *The Australasian Medical Gazette*, p. 296, 20 July 1896.
16. Harris H., 'Notes on the Roentgen Rays and Radium: A Digest of Sixteen Years Experience and Over Thirty Thousand Cases', *The Australasian Medical Gazette*, p. 197-206, 7 Mar. 1914.
17. Radford E.P., Royal Commission into British Nuclear Tests in Australia, *Transcript*, p. 4756, AGPS, 1985. (Hereafter cited as Transcript.)
18. Stewart A., Webb J. & Hewitt D., 'A Survey of Childhood Malignancies', *British Medical Journal*, pp. 1495-1508, 1958.
19. Stewart A. & Kneale G., 'Radiation Dose Effects in Relation to Obstetrics, X-ray and Child Cancer', *Lancet*, pp. 1185-7, 1970.
20. MacMahon B., 'Prenatal X-ray Exposure and Childhood Cancer', *Journal National Cancer Institute*, Vol. 28, p. 1173, 1962.
21. Harvey E.B., Boyce J.D., Honeyman M. & Flannery J.T., 'Prenatal X-ray Exposure and Childhood Cancers in Twins', *New England Journal of Medicine*, Vol. 312, pp. 541-5, 1985.
22. Green P., 'The response of the International Commission on Radiological Protection to calls for a reduction in the dose limits for radiation workers and members of the public', *International Journal of Radiation Biology*, Vol. 53, pp. 679-82, 1988.
23. Moore R.M., Barrick M.K. & Hamilton P.M., 'Effects of Sonic Radiation on Growth and Development', *American Journal Epidemiology*, Vol. 116, p. 571 (abstract), 1982.
24. World Health Organisation, Environmental Health Criteria 22, *Ultrasound*, 1982. See also Repacholi, M.H., 'Medical Ultrasound: Is There a Hazard?' *Australian Physical & Engineering Science*, Vol. 6, pp. 64-7, 1983.
25. Radford E.P., *Transcript*, pp. 4748-9.
26. Bertell R., 'The nuclear worker and ionising radiation', *American Industrial Hygiene Journal*, Vol. 40, pp. 395-401, 1979.
27. 'A nun fights nuclear horrors', *Age*, 20 August 1985.
28. Bross I.D. & Natarajan N., 'Genetic Damage from Diagnostic Radiation', *Journal American Medical Association*, Vol. 237, p. 2399, 1977. See also by the same authors 'Cumulative genetic damage in children exposed to preconception and intrauterine radiation', *Investigative Radiology*, Vol. 15, pp. 231-45, 1980.
29. Subcommittee on Health and the Environment, US House of Representatives, *Effect of Radiation on Human Health*, Vol. 1, pp. 905-1000, Serial No. 95-179, Government Printing Office, Washington DC, 1979.

See also 'Locking Up Health Data' in Chapter 10 which deals with the Mancuso Affair.

30. Morris N.D., *An Examination of the Distribution of Patient Doses from Diagnostic X-ray Procedures*, Australian Radiation Laboratory Report, TR051, pp. 7-8, 1983.
31. For example the Health *(Radiation Safety) Act*, was passed by the Victorian Government in 1983 and the Health *(Radiation Safety Regulations)* approved in May, 1984 with amendments to those rules in July 1984.
32. Wright J., 'X-rays A Mixed Blessing', *New Doctor*, Issue 49, p. 18-20, Nov., 1988.
33. World Health Organisation Scientific Group, *A Rational Approach to Radiodiagnostic Investigations*, p. 7, Geneva, 1983.
34. 'Fear of legal redress prompts many unnecessary X-rays, doctor warns', *Age*, 16 Feb. 1987.
35. See *Ref.* 32.
36. See *Ref.* 12, Gofman & O'Connor, Ch.16 'Low Cost Ways to Reduce Dose and Risk', pp. 347-64, Ch.17 'Assessing the Aggregate Cancers from Diagnostic X-rays', 369-73.
37. 'Dentists in Medi Scram', *Sun*, 24 Dec. 1985.
38. See *Ref.* 33, p. 8.

Chapter 2

1. Harris H., 'Notes on the Roentgen Rays and Radium: A Digest of Sixteen Years Experience and Over Thirty Thousand Cases', *The Australasian Medical Gazette*, p. 197-206, 7 Mar. 1914.
2. Steneck N.H., Cook, H.J., Vander, A.J. & Kane G.L., 'The Origins of US Safety Standards for Microwave Radiation', *Science*, Vol. 208, pp. 1230-7, 1980.
3. Becker R.O. & Marino A.A., *Electromagnetism and Life*, State University of New York Press, 1982, Chapters 5 and 6 discuss effects of EMR on nervous and endocrinal systems. See also Baranski, S. & Czerski, P., *Biological Effects of Microwaves*, Dowden, Hutchinson & Ross, Stroudsburg, Penn., 1976.
4. Battocletti J.H., *Electromagnetism, Man and the Environment*, p. 68, Elek Books, London, 1976.
5. *Consumer Reports*, pp. 221-30, Mar. 1973.
6. *Choice*, p. 166-72, June 1979.
7. Glazer Z.R. & Dodge C.H., 'Comments on occupational safety and health practices in the USSR and some Eastern European countries: Possible dilemma in risk assessment of RF and Microwave radiation bioeffects', *Risk Benefit Analysis — The Microwave Case*, Ed. N. Steneck, pp. 53-67, San Francisco Press, 1982.
8. Brodeur B., *The Zapping of America — Microwaves, Their Deadly Risk and The Cover-up*, Norton, NY, 1977.

9. *Ibid*, Brodeur sees the Moscow affair as a conspiracy to suppress information because the US defence authorities feared public opposition to microwave installations. Steneck N.H. in *The Microwave Debate*, pp. 92-118; MIT Press, Cambridge, Mass., 1984, takes the view that it was more complicated and involved not only the interests of the military but also conflicting motivations within the scientific community. Steneck *et al.* expand on this theme in an account of the setting of RF safety standards in *Ref.* 2. See also Steneck N.H., 'The relationship of history to policy', *Science, Technology & Human Values*, Vol. 7, pp. 105-12, Summer 1982.
10. US State Department Memorandum, 13 Feb. 1967.
11. Cesaro R.S., US State Department Memorandum, 20 Dec. 1966.
12. Johnston J., US State Department Aid Memoir, 25 Sept. 1988.

Chapter 3

1. Steneck N.H., *The Microwave Debate*, MIT Press, Cambridge, Mass., 1984. Steneck gives an account of the Rockaway inquiry in Ch. 1, 'The Skeptical Public'.
2. *Emerald Hill Times*, 12 Sept. 1985.
3. *Open-Mike*, ABC Radio, Melbourne, 1 Aug. 1985.
4. *Earthworm*, ABC Radio, 13 May 1987.
5. *Ibid.*
6. *Ibid.*
7. Australian Standard 2772, *Maximum Exposure Levels — Radio-Frequency Radiation — 300kHz to 300GHz*, Standards Association of Australia, 1985.
8. Steneck N., Cook H.J., Vander A.J. & Kane G.L., 'The Origins of US Safety Standards for Microwave Radiation', *Science*, Vol. 208, pp. 1230-7, 1980. An analysis of the mixed motivations of scientists working on developing a scientific basis for the 1974 US RF standard.
9. See *Ref.* 7, p. 4.
10. Hollway D.L., 'The Australian Safety Standard for RF Radiation — A Curate's Egg', Division of Applied Physics, CSIRO, 1985.
11. *Technology Report*, ABC, 'The Australian Electromagnetic Radiation Standard', March 1985.
12. See *Ref.* 10.
13. Vosburgh B., *Proceedings of the Tri-Service Conference on the Biological Hazards of Microwave Radiation*, p. 120, 1958; quoted in *Ref.* 1, p. 53.
14. *Technology Report*, ABC, 'Setting the Australian EMR Standard'. Part 2.
15. *Science Show*, ABC, 'Electromagnetic Radiation', Program 3, May 1984.
16. Medici R. G., 'Considerations For Science. Where Has All The Science Gone?', *Risk/Benefit Analysis. The Microwave Case*, edited by N. Steneck, p. 117, San Francisco Press, 1982.

17. *Op cit,*. p. 179.
18. *Microwave News*, Vol. IV, No. 6, July/August 1984.
19. *Ibid*.
20. Prausnitz S. & Susskind C., 'Effects of Chronic Microwave Irradiation in Mice', *IRE Transactions on Biomedical Electronics*, Vol. 9, p. 104, 1962.

Chapter 4

1. Becker R.O., 'Currently at Risk, *The Good, The Bad and The Indefensible,* TV Channel 4, UK, 1984.
2. American Institute of Biological Sciences, *Biological and Human Health Effects of Extremely Low Frequency Electromagnetic Fields. Post 1977. Literature Review,* Report of the Committee on Biological Effects of Extremely Low Frequency Electromagnetic Fields, Arlington, Va, USA, 1985.
3. Wisconsin Justice Department, 'Commentary on the American Institute of Biological Sciences Literature Review, Biological and Human Health Effects of Extremely Low Level Electromagnetic Fields', July 1985.
4. State Electricity Commission of Victoria, *Environment Effects Statement: Proposed 220kV Transmission Line, Brunswick-Richmond*: Part A: 'A Review of Environmental Issues'. Part B: 'Biological and Human Health Effects from Transmission Lines and Stations in Victoria', Nov. 1985.
5. ABC TV Channel 2, *Four-Corners* Transcripts, 'Low-Level Electromagnetic Radiation', June 1985; 'Powerlines and Radiation', Sept. 1985, and 'Electric Power Lines: A Follow-Up Report', June 1986; Parliamentary Library, Canberra.
6. '$25 million jury verdict against HL&P Ruling supports claim that powerline could be harmful', *Houston Post*, 28 Nov. 1985.
7. See *Ref* 4, Part A, pp. 12-13.
8. Smith J., General Manager, SECV, *The Age,* 20 Jan. 1984.
9. See *Ref.* 4, Part B, p. 66.
10. The policy of the Collingwood Residents Association (CRA) endorsed at public meetings was that 'a public enquiry with community representation be held into the health effects of existing and proposed high-voltage powerlines'. The CRA *Submission on the SECV's Environment Effects Statement for the Proposed Brunswick to Richmond 220KV Transmission Line*, says about the planning process '. . . it is unreasonable the community is forced to interpret and respond to the issue through the restrictive EES procedures as well as rely on its own limited resources to mount a defence.'
11. Personal communication to author from Victorian Minister of Health, David White, 11 Dec. 1985.
12. Bates I., Chief Engineer State Electricity Commission of Victoria, in

a letter to K.H. Lokan, NHMRC, 7 Oct. 1985., *SECV ref. no. FOI 86/59*, 30 July 1986. Obtained under Victorian Freedom of Information Act.

13. See *Ref.* 5, June 1986.
14. Communication between lay and scientific communities is illustrated by *Electromagnetic Fields, Electric Power and Public Health — A Community Resource Document Based on the Victorian Experience*, by Ian Macmillan on behalf of Collingwood Residents Association and Collingwood Health Centre, Victoria 1986. The report lists correspondence between the residents and leading researchers in non-ionising EMR. There is now quite a number of regular community environmental and energy bulletins that circulate internationally.
15. Graves H.B., *A Review of the State Electricity Commission of Victoria (SECV) Transmission Line and Station Design Practices in Relationship to the Health and Safety of People*, prepared for Minister for Health, Victorian Government, Australia, March 1986.
16. Moore J., *Powerlines*, video of public meeting between H.B. Graves and Rangebank school parents. Available from Collingwood Community Health Centre, Sackville St, Collingwood, Victoria.
17. The Powerline Review Panel was presided over by David Scott, Commissioner For the Environment. The panel first issued a public consultation document on *Information Options and Outlooks* before public hearings with government and community organisations. Following public hearings, draft recommendations were made to the state government for comment and public consultative sessions were called. The panel then prepared its final report to the state government. Among the reports commissioned by the panel were: *A Critical Examination of the Brunswick to Richmond Powerline Dispute, 1968-1987*, by Trevor Blake, *Recreation. Amenity Values and Metropolitan Powerline Corridors — Issues, Approaches Sources* by David Mercer and *Environmental Contaminants as Public Health Risks with Particular Reference to Electric Transmission Lines*, by S. Martin Taylor.
18. The Powerline Review Panel, *Final Report to the Victorian Government*, Appendix iii, p. ii, July 1989.
19. See *Ref.* 5.
20. Young L., 'Electromagnetic pollution, a new field of study', *Age*, 21 Sept. 1987.
21. Wertheimer N. & Leeper E.,. 'Electrical Wiring Configuration and Childhood Cancer', *American Journal of Epidemiology*, Vol. 109, pp. 273-84, 1979. 'Adult Cancer Related to Electrical Wires Near the Home', *International Journal of Epidemiology*, Vol. 11, pp. 345-55, 1982.
22. *Microwave News*, p. 6., July/Aug. 1986.
23. Delpizzo L., 'Wire coding as an indicator of residential magnetic field exposure' *Carcinogenic Potential of Extremely Low Frequency Magnetic Fields*, Workshop Proceedings, pp. 104-16, Australian Radiation Laboratory, 1988.

24. See *Ref.* 4, Part A, p. 124. For a discussion of the issue of residential magnetic fields see I. McMillan, *Ref.* 13, Section 6, pp. 3-38.
25. Tomenius L., '50 Hz Electromagnetic Environment and the Incidence of Childhood Tumors in Stockholm County', *Bioelectromagnetics*, Vol. 7, pp. 191-207, 1986.
26. Mild K., *et al.*, 'Effect of High-Voltage Pulses on the Viability of Human Leucocytes *In Vitro, Bioelectromagnetics*, Vol. 3, p. 213, 1982; Wright E.W., *et al.*, 'Leukaemia in Workers Exposed to Electrical and Magnetic Fields,' *Lancet* p. 1160, 20 Nov. 1982; Milham S., 'Mortality from Leukaemia in Workers Exposed to Electrical and Magnetic Fields', *Lancet*, p. 249, 22 July 1982; Pearce N.E. *et al.*, 'Leukaemia in Electrical Workers in New Zealand,' *Lancet*, p. 811, 6 April 1985; 'Radio Operators,' *Lancet*, p. 812, 6 April 1985; Seager J., 'Powerline Workers Contracting Leukemia', *Journal National Cancer Institute*, Vol. 70, p. 37-44, 1985; Bauchinger M., *et al.*, 'Analysis of Structural Chromosome Changes and SCE after Occupational Long-term Exposure to Electric and Magnetic Fields from 380 kV Systems', *Radiation and Environmental Biophysics*, Vol. 19, pp. 235-8, 1981; Lin R.S., *et al.*, 'Occupational Exposure to Electromagnetic Fields and the Occurrence of Brain Tumors', *Journal of Occupational Medicine*, Vol. 27, pp. 413-19, 1985; Speers M. *et al.*, *Journal Industrial Medicine*, 1985.
27. Fulton J.P. *et al.*, 'Electrical Wiring Configurations and Childhood Leukemia in Rhode Island,' *American Journal of Epidemiology*, Vol. 3, pp. 293-6, 1980; McDowall M.E., 'Leukaemia Mortality in Electrical Workers in England and Wales', *Lancet*, p. 246, 29 Jan. 1983; Coleman M., *Microwave News*, Mar./April 1986; McDowall M.E., 'Mortality of persons resident in the vicinity of electrical transmission facilities', *British Journal of Cancer*, Vol. 53, pp. 271-9, 1986. For summary of studies of cancer incidence see *Ref.* 23, Repacholi, M.H., Workshop Proceedings, pp. 1-16.
28. *Biological Effects of Powerline Fields*, New York State Powerlines Project, Scientific Advisory Panel Final Report, p. 10, 1987.
29. Savitz D., Wachtel F., Barnes F.A., John E.M. and Tvrdik, J.G., 'Case-Control Study of Childhood Cancer and Exposure to 60Hz Magnetic Fields', *American Journal of Epidemiology*, Vol. 128, pp. 21-38, 1988.
30. See *Ref.* 20.
31. *Electromagnetic Fields from Overhead Transmission Lines and Underground Cables*, Report to the Victorian Department of Technology, Energy and Resources by a panel chaired by Professor W.J. Bonwick, May 1988.
32. Armstrong B., *A Review of Epidemiological Data on the Effects of Extremely-Low-Frequency Electromagnetic Fields (ELF) on Human Health*, Department of Medicine, University of Western Australia, December 1986.
33. Repacholi M.H., 'ELF and cancer: an overview', see *Ref.* 21, pp. 1-16.
34. Delpizzo V., 'An Evaluation of the Existing Evidence on the Carcino-

genic Potential of Extremely Low Frequency Magnetic Fields', *Australasian Physical & Engineering Sciences in Medicine*, pp. 55-68, Vol. 12, No. 2, pp. 55-68, 1989.
35. *Energy Forum*, p. 7, Victorian Department of Industry, Technology and Resources, Dec. 1989.
36. 'International Utility Symposium on the Health Effects of Electric and Magnetic Fields,' Toronto, Sept. 1986. *Microwave News*, p. 1, Sept./Oct. 1986. See also *Ref.* 13 for excerpts from transcripts showing where the experts differed on observed bioeffects.

Chapter 5

1. *Science Show*, ABC, 'Electromagnetic Radiation', Program 3, May 1984.
2. *Ibid.*
3. Joyner K.H. & Bangay M.J., *RF Dielectric Heaters. Survey Results and Guidelines for Limiting Occupational Exposures to Radiofrequency Radiation*, Australian Radiation Laboratory, Report TR052, 1983.
4. Australian Standards Association, Drafting Committee on Microwave and Radiofrequency Radiation Hazards, *Minutes*, 28 Nov. 1979.
5. Mathews J., *Health and Safety at Work — Australian Trade Union Safety Representatives Handbook*, p. 173, Pluto Press, Sydney, 1985.
6. See *Ref.* 1.
7. Brown L.J. versus Commonwealth of Australia, Federal Administrative Tribunal, 'Compensation — eye cataract — whether attributed to microwave radiation at work — conflicting medical evidence', 19 April 1985.
8. *Op.cit.*, p. 140.
9. *Op.cit.*, p. 125.
10. *Op.cit.*, p. 79.
11. *Op.cit.*, p. 116.
12. *Op.cit.*, 'Reasons for Decision', p. 11.
13. Steneck N.H., *The Microwave Debate*, p. 179, MIT Press, Cambridge, Mass., 1984.
14. *Op.cit.*, p. 16.
15. Zaret M. & Snyder, W., 'Cataracts and Avionic Radiations', *British Journal Ophthalmology*, Vol. 61, p. 383, 1977.
16. Hollows F.C. & Douglas J.B., 'Microwave Cataract in Radiolinemen and Controls', *Lancet*, p. 406, 18 Aug. 1984.
17. *Technology Report*, ABC, 'Electromagnetic Radiation and Cataract', 14 Nov. 1984.
18. Foard G., *In The Face of Uncertainty — A Report on Radiation Standards*, p. iv, Community Education Publication Association, Carlton, Victoria, Australia.
19. Joyner K.H. *et al.*, *Electromagnetic Emissions from Video Display Terminals (VDTs)*, Australian Radiation Laboratory, Report ARL/TR067, 1984.

20. Swedish Institute of Radiation Protection, *Radiation from VDTs*, Report: a84 — 08, Stockholm, 1984. (In Swedish.)
21. Kivisakk E., Non-Ionising Radiation Section, Swedish Institute of Radiation Protection, Stockholm, personal communication, July 1988. See also Tribukait B., Cekan E. and Paulsson L-E., 'Effects of pulsed magnetic fields on embryonic development in mice', *Work With Display Units 86*, Elsevier, Netherlands, 1987 and Frolen H., Svedenstal B-M, Bierke P., and Fellner-Feldegg H., 'Teratogenic Effects of Pulsed Magnetic Fields', Pathology Institute, University of Sveriges Lantbruks, Sweden. (In Swedish with English summary.)
22. 'Henhouse Project: Weak PMFs Cause Chick Abnormalities', *Microwave News*, p1, 14, Mar./April 1988.
23. *Microwave News*, July/August 1986.
24. 'VDT Use Linked to an Increased Miscarriage Risk', *Microwave News*, p1, 13, May/June 1988.
25. See *Ref.* 19, p. 15.
26. Personal Communication from G. Elliot, Non-Ionising Radiation Group, Australian Radiation Laboratory, 23 Jan. 1985. Questions concerning the significance of pulsing emissions posed by Swedish scientists were left unanswered.
27. 'Perils of electronic radiation', *Far Eastern Economic Review*, p. 74, 22 Feb. 1990.
28. Brodeur P., 'Annals of Radiation — The Hazards of Electromagnetic Fields', Part III 'Video Display Terminals', *The New Yorker*, p. 68, 26 June 1989.
29. Hollway D.L., 'The Australian Safety Standard for RF Radiation — A Curate's Egg', Division of Applied Physics, Commonwealth Scientific and Industrial Research Organisation, 1985. For the original quotation see Steneck N.H., 'Subjectivity in standards: the case of the ANSI C95.1-1982', *Microwaves and RF*, pp. 137-9, 141-2, 164-6, May 1983.

PART II

Chapter 6

1. Badash L., 'Radium, radioactivity and the popularity of scientific discovery', *Proceedings of the American Philosophical Society*, Vol. 122, pp. 145-54, 1978.
2. Lawrence H.L., 'Radium and Electro-Therapies in the Treatment of Diseases of the Skin', *Transactions 7th Session Australasian Medical Congress*, 1905, pp. 44-8.
3. Richardson J.F., *Australian Radiation Laboratory — A Concise History, 1929-1979*, AGPS, 1981. The inventory of radium in Australia up to 1927 was 715 mg for private practice, 145 mg. for hospitals and an unspecified quantity held by surgical firms which they hired to

'approved' persons. See also Hamersley H., 'Radiation Science and Australian Medicine — 1896-1914,' *Historical Records of Australian Science*, Australian Academy of Sciences, Canberra, 1981.

4. The saga of radioactive waste dumped at Hunters Hill has lasted more than a decade. The Environmental and Special Health Services, Health Commission of New South Wales prepared many reports on disposing of the waste. Surveys by the commission showed heavy ground contamination. The proposed solution was temporary storage until a permanent resting place could be found. The favoured site was in the far west of the state but this met fierce regional resistance. The outcome has been a hefty Health Commission file but no action apart from evacuating houses. See 'Cabinet to consider waste report', *Sydney Morning Herald*, 17 Jan. 1978; '500 tons treated at radium factory', *Daily Telegraph*, 22 Jan. 1978; 'Two sites to dump radioactive soil', Daily Telegraph, 5 Nov. 1980; 'Don't dump radium out here', *Sydney Morning Herald*, 25 Jan. 1978; '$1m. for Hunters Hill clearance', *Financial Review*, 28 Nov., 1980. Other dumps, apart from uranium mine sites, are at Byron Bay in NSW and Capel in Western Australia (beach sand wastes), Bairnsdale in Victoria, and Adelaide in South Australia.
5. Radford E.P., *Transcript*, pp. 4756-9, 1985. Radford was called by the Royal Commission as an expert on low-level ionising radiation. He heads the re-evaluation of the Hiroshima data.
6. Caufield C., 'The Dial Painters', *Multiple Exposures*, pp. 38-40, Secker & Warburg, London, 1989.
7. Spiess H. & Mays C.W., 'Bone Cancers Induced by Radium-224 in Children and Adults', *Health Physics*, Vol. 19, pp. 713-29.
8. 'Regulations Regarding The Storage, Mixing, Manipulating or Use in Factories of Radioactive Paints', *Victorian Government Gazette*, December 1944.
9. See *Ref.* 1, p. 147.
10. Rowntree L.G. & Baetjer W.A., *Journal American Medical Association*, p. 1442, 1913.
11. *Ibid.*
12. Schubert J. & Lapp R.E., *Radiation — What It Is and How It Affects You*; Heinemann, 1957. See Chapter 6, 'Radioactive Poisons', for an account of early misuse of radium.
13. See Hamersley *Ref.* 3
14. Lawrence H.F., *Radium: when and how to use it*, p. 108, Stillwell, Melbourne, 1911.
15. Oliphant M., *10th Annual Conference*, Australian Radiation Protection Association, 1982.
16. World Health Organisation, 'Radiological Examination of Drinking Waters', Report of a Working Group, *EURO Report No. 17*, 1979.
17. Gofman J.W., *Radiation and Human Health*, pp. 438-9; Sierra Club Books, San Francisco, 1981.
18. Cooper M.B., Ralph B.J. & Wilkes M.J., *Natural Radioactivity in Bottled*

Mineral Water available in Australia, Australian Radiation Laboratory, Report TR 036, 1981.
19. See *Ref.* 17, p. 439.

Chapter 7

1. Rhodes R., *The Making of the Atomic Bomb*, p. 649, Simon & Schuster, London, 1986. In the Chapter 'Trinity' Rhodes describes the debate on the use of the bomb between scientists and administrators. Politicians and scientists began to rationalise nuclear war using innuendo rather than face the moral issue. This aspect is more thoroughly explored by Gar Alperovitz, in his introduction to the 1985 edition of his *Atomic Diplomacy* (Penguin). Surprisingly the Chief of Staff, Admiral Leahy, opposed the dropping of the bombs whereas leading scientists and politicians thought there was 'no argument' about dropping them.
2. *Op.cit.*, pp. 324-5. In 1940 Frisch and Peierls reported to the British Government that the critical mass of uranium-235 was small enough to make an A-bomb practical. In an accompanying document *Memorandum on the properties of a radioactive 'superbomb'* they warned on the dangers of radioactivity.
3. Committee for the Compilation of Materials on Damage by the Atomic Bombings on Hiroshima and Nagasaki, *Hiroshima and Nagasaki — The Physical, Medical and Social Effects of the Atomic Bombings*, p. 130-85, Hutchinson, London, 1981. For a description of the black rain see p. 92.
4. Goodchild P., *J. Robert Oppenheimer — 'Shatterer of Worlds'*, p. 171, British Broadcasting Corporation, London, 1980.
5. See *Ref.* 3, pp. 136-331. See also Radford E., 'Recent evidence of radiation-induced cancer in the Japanese atomic bomb survivors', *Radiation and Health — The biological effects of low-level exposure to ionizing radiation*, pp. 76-96, Ed. Russell Jones R. & Southwood R., Wiley, Brisbane, 1987.
6. See *Ref.* 3, p. 222-35.
7. Sennett J., *Hiroshima — A Personal Account*, a video, Movement Against Uranium Mining (Vic), Melbourne. See Stephen Kelen, who was also in the Australian Occupation Forces, *I Remember Hiroshima*, Hale & Iremonger, Sydney, 1983. Other books dealing with the medical and social effects of the bomb's radiation are: Mabusa, Masuji. *Black Rain*, John Martin, London, 1969; Townsend, Peter, *The Postman of Nagasaki*, Penguin, 1984; *Unforgettable Fire — Pictures Drawn by the Atomic Bomb Survivors*, Japanese Broadcasting Corporation, Wildwood House, London, 1981.
8. Pilger J., *Heroes*, p. 494, Jonathan Cape, London, 1986.
9. *New York Times*, 13 Sept. 1945. Since the A-bombs exploded about 600 metres above ground the radioactivity was not as great as it would have been from ground explosions. However, significant radioactivity

remained for days; some of the early entrants to the cities, who came to rescue victims, were later victims of the residual radioactivity. See 'Induced Radiation', *Ref.* 3, pp. 74-9.
10. See *Ref.* 3, p. 15.
11. Rotblat J., 'A tale of two cities', *New Scientist*, pp. 46-50, 7 Jan. 1988. An account of the Life Span Study and of the debate over their meaning for radiation risks. See also Radford E., *Ref.* 5.

Chapter 8

1. Royal Commission into British Nuclear Tests in Australia, *Transcript*, p. 4443, Australian Government Publishing Service, Canberra, 1985. (Hereafter cited as *Transcript* or *Exhibit.*)
2. Johnstone D.R., 'Conditions at Maralinga' statement in the case of Daryl Johnstone v. The Commonwealth of Australia, Supreme Court of NSW, 1980. See also Johnstone's testimony to the Royal Commission into British Nuclear Tests in Australia. Johnstone was awarded small compensation. He continued his legal action until in 1988 he was awarded much larger compensation.
3. Johnstone D.R., *Transcript*, p. 193; also *Exhibit* RC18.
4. See *Ref* 2, Statement to Supreme Court.
5. *Ibid.*
6. Evans J., Expert evidence to the Supreme Court in the Johnstone case.
7. Bird C., *Exhibit* RC46.
8. Accounts of experiences of aircrews sent on air-sampling missions: Wilkes, J. *Field of Thunder*, p. 31, Friends of the Earth, Melbourne, 1981; Blakeway D. & Lloyd-Roberts S., *Fields of Thunder — Testing Britain's Bomb*, Ch. 6, Unwin Paperbacks, 1985; Smith J., *Clouds of Deceit — The Deadly Legacy of Britain's Bomb Tests*, p. 106, Faber & Faber, 1985.
9. Milliken R., *No Conceivable Injury — The story of Britain and Australia's cover-up*, pp. 160-1, Penguin, 1986. The statement is contained in a document of the British Defence Research Policy Committee. Penney told the Royal Commission he thought the document 'pretty dreadful because it can lead to misunderstandings.' See also Lord Penney, *Transcript*, p. 4440.
10. Lord Penney, *Transcript*, p. 4442.
11. *Transcript*, p. 4443
12. *Exhibit* RC328. Ministerial submission to Federal Cabinet, 4 Sept. 1956; handout to newspaper editors *Exhibit* AG9.
13. Royal Commission into British Nuclear Tests in Australia, *Conclusions and Recommendations*, p. 30, para. 201, AGPS, 1985. (Hereafter cited as *Conclusions and Recommendations*).
14. Quoted by Blakeway & Lloyd-Roberts, *Ref.* 8, p. 131.
15. *Conclusions and Recommendations*, p. 30, para.24.
16. Lord Penney, *Transcript*, p. 4441.

17. Titterton E.W., 'Why Australia is an Atom Testing Ground', *Age*, 16 May 1956.
18. 'Buffalo Trials Indoctrinee Instruction No. 1', attached to statement by F.S.B. Peach, *Exhibit* RC 541.
19. 'Atom Survivors Lay Siege to a Political Wall of Secrecy', *Age*, 17 March 1984.
20. See *Ref.* 18.
21. Wasserman H. & Solomon N., *Killing Our Own*, Part 1: 'The Bombs,' pp. 3-57; Delacorte Press, 1982, describes the struggles of US nuclear veterans for compensation. For Australian veterans' campaign see Tame A. & Robotham F.P.J., *Maralinga*, 'Untruths and Half-Truths,' pp. 178-96, Fontana, Melbourne, 1982.
22. Dannielsson B. & M.T., *Poisoned Reign — French Nuclear Colonialism in the Pacific*, p. 284; Penguin, Melbourne, 1986.
23. Carrick J.L., Minister for National Development and Energy, Statement to the Senate, 15 May 1980.
24. Australian Ionising Radiation Advisory Council, *British Nuclear Tests in Australia — A review of operational safety measures and of possible after-effects*, AIRAC Report No. 9: AGPS, 1983.
25. Kerr C.B., Bently K., Keam D.W. and Robotham R.P.J., *Report on the Expert Committee on the Review of Data on Atmospheric Fallout Arising from British Tests in Australia*, AGPS, 1984.
26. Letter from the Australian Ionising Radiation Advisory Council to Barry Cohen, Minister for Home Affairs and Environment, 13 August 1984; *Exhibit* RC 67.
27. *Conclusions and Recommendations* p. 29, para. 197.
28. *Op.cit.*, p. 15, para. 74.
29. 'Maralinga Man Says Death Is Official Policy', *Age*, 4 April 1988.
30. *Conclusions and Recommendations*, para. 51, p. 12.
31. Green P.A., *International Commission on Radiological Protection*, Greenpeace, London.
32. Green P.A., 'ICRP — Gentleman's Club', *SCRAM Journal*, p. 12, July/Aug. 1987.
33. Radford E.P., 'Statement concerning proposed Federal radiation protection guidance for occupational exposures', presented originally to the US Environmental Protection Agency Office of Radiations Programs, *Exhibit* RC259.
34. Royal Commission into British Nuclear Tests in Australia, *Report*, Vol. 1, pp. 97-8, AGPS, 1985. (Hereafter cited as *Report*).

Chapter 9

1. Lapp R.E., *The Voyage of the Lucky Dragon*, Penguin, 1958. Lapp relates the story as told to him by the Japanese fishermen. See also Schubert J. & Lapp R.E., *Radiation — What it is and how it affects you*, Ch. 1: 'The Awesome Cloud,' Heinemann, London, 1957.

2. *Marshall Islands — A Chronology 1944-81*, p. 10. Also Johnson G., *Collision Course at Kwajalein — Marshall Islanders in the Shadow of the Bomb.* Both publications are available from Pacific Concerns Resource Centre, P.O. Box 9295, Newmarket, Aukland, New Zealand.
3. *New York Times*, 20 Sept. 1982.
4. The story of the Rongelap Islanders and US weather technicians in the wake of the fallout is documented in the film *Half Life* by O'Rourke & Associates Filmmakers Pty Ltd, GPO Box 199, Canberra.
5. Conard R.A. *et al*, 'Medical Survey of Rongelap and Uterik People Three Years After Exposure to Radioactive Fallout, *Report BNL 501*, p. 22; Brookhaven National Laboratory, June 1958.
6. Teller E. & Latter A.L., *Our Nuclear Future*, p. 94, 1958. Quoted in *Defended Unto Death*, Gwyn Prins (Ed.), p. 75, Penguin, 1983.
7. Costeau J., 'Tears in the Ice-cream'. Quoted in Prins, G. (Ed.), *Defended Unto Death*, p. 251-2; Penguin, 1983.
8. Titterton E., 'Why Australia is an Atom Testing Ground,' *Age*, p. 2, 19 May 1956.
9. *Exhibit* AG 9, pp. 897-911.
10. Cable from Lord Penney at Maralinga to AWRE Aldermaston, *Exhibit* AG 10.
11. 'Studies Confirm Massive Increase in Fallout Over Melbourne', *The Age*, 7 May 1984.
12. 'Bomb Test Town Has a Sad Legacy,' *Age*, p. 11, 14 March 1984. See also Ball H., *Justice Downwind — America's Atomic Testing Program in the 1950s*', Ch.4 'The Downwinders and the Trauma of Cancer,' pp. 84-101, Oxford University Press, NY, 1986.
13. Johnson C., *Journal American Medical Association*, p. 251, Jan. 1984.
14. Irene Allen, *et al.*, Plaintiffs v. United States of America in the District Court of Utah, Central Division, United States District Judge, 9 May 1984. See Ball in *Ref.* 12, pp. 145-69.
15. Eames G., *Transcript*, pp. 7083-6.
16. Lester Y., *Transcript*, p. 7117.
17. Kanytji, *Transcript*, pp. 7189-92.
18. Lester Y., *Transcript*, p. 7131.
19. *AM Public Affairs*, ABC Radio, 14 May 1980, *Exhibit* RC 800.
20. Lester Y., *Transcript*, p. 7132.
21. 'High Explosives Research Report', AWRE No. A32, May 1953, *Exhibit* RC247.
22. 'Transport of Debris From the British Nuclear Test in Sth Australia on 15 Oct. 1953', AWRE Tech. Div. Note 8/84, Aug. 1984, *Exhibit*, RC253.
23. *Transcript*, pp. 9472-6. The cross-questioning of Sir E. Pochin, British National Radiation Protection Board, pp. 9329-476 provides insights into the uncertainties of radiation protection. Variation in susceptibility of individuals to low-level radiation according to age and other factors is also dealt with at length during questioning of E. Radford, *Transcript*, pp. 4738-806 and A. Stewart, pp. 6717-61.

24. *Report*, Vol. 1, p. 190.
25. *Ibid*. p. 194.
26. Lander A., *Transcript*, p. 7098.
27. 'Radioactive Rain Reported in North', *Age*, p. 1, 23 June 1956; also Dwyer's notes on Mosaic II, *Exhibit* RC555.
28. *Conclusions and Recommendations*, para.116, p. 19.
29. Fadden's public utterances and the government's private reactions to the cloud passing over the mainland are described in 'Ill Wind at Monte Bello', *Australian*, 23 June 1984. The Royal Commission failed to locate the message but Bernard Perkins a radio operator on the HMS *Narvik* says that he was on duty when the signal was received and written down.
30. *News* (Adelaide), 21 June 1956; see Tame & Robotham, *Maralinga*, Fontana, Melbourne, 1982 for Marston's reactions to government assurances on Mosaic test fallout.
31. Hamersley H. & Moroney J., 'Hedley R. Marston, FRS and the Atomic Weapons Safety Committee — Controversy over fallout from British Nuclear Tests in Australia in 1956', a paper prepared for the Royal Commission into British Nuclear Tests in Australia, p. 42. See also Tame A. & Robotham F.P.J., *Ref* 30, pp. 116-30, and Blakeway & Lloyd-Roberts, *Fields of Thunder — Testing Britain's Bombs*, p. 137; Unwin Paperbacks, Sydney, 1985.
32. Hamersley & Moroney, *op.cit.*, pp. 23-4.
33. Lord Penney, *Transcript*, p. 4407. A document titled 'Indoctrination of the Public' was discussed at a meeting of the Safety Committee on 28 July 1956. The minutes of the meeting record that, 'The members appreciated the need for indoctrination of the public and that they would check any official release submitted to the committee'.
34. Pauling L., *No More Wars*, p. 74-5; Dodd, Mead, 1958.
35. Pauling L., 'How Dangerous is Radioactive Fallout?' *Foreign Affairs Bulletin*, June, 1957 p. 149.
36. Bertell R., *No Immediate Danger — Prognosis for a Radioactive Earth*, p. 105; The Women's Press, London, 1985.
37. From speech by Chief Seattle, leader of the Suquamish Tribe, marking the transfer of the territory now Washington State to the Government in 1854, *Active Non-Violence in the United States, The Power of the People*. From Peace Press Inc., 3828 Willat Ave, Culver City, Cal., USA.

Chapter 10

1. De Laguna W., 'What is Safe Waste Disposal?', *The Bulletin of Atomic Scientists*, Vol. 15, p. 35, March 1959.
2. Butler A.M., Keys F.G. & Szent-Gyorgy A., *The Bulletin of Atomic Scientists*, Vol. 16, p. 141, April 1960.
3. Quoted by Falk, J., 'Origins of the Opposition', *Global Fission — The Battle over Nuclear Power*, p. 95, 1982.

4. *The Four Corners — A National Sacrifice Area*, 1983, a documentary film available from Friends of the Earth, Melbourne.
5. *Uranium in Our Backyard*, Bullfrog Films, USA.
6. See *Ref.* 4.
7. Shapiro F.C., 'A Reporter at Large Nuclear Waste', *New Yorker*, pp. 53-140, 19 Oct. 1981.
8. ABC TV Channel 2, *Quantum*, 'Kakadu', 1989.
9. See *Ref.* 4; also US National Academy of Science, *Rehabilitation of Western Coal Lands*, pp. 85-6, Ballinger, Cambridge Mass., 1974.
10. 'Massive plutonium levels found in Cumbrian corpses', *New Scientist*, p. 11, 14 Aug. 1986.
11. *Windscale [Sellafield] Nuclear Reprocessing Plant in Britain*, Yorkshire Television, UK, 1984.
12. Urquart J., *Leukaemia and Nuclear Power in Britain — The Evidence so Far*, Friends of the Earth, London, UK. Martin, S. 'Cluster & Super Clusters', *SCRAM Journal*, pp. 12-13, Jan./Feb. 1987.
13. *Lancet*, Aug. 1984; *Nature*, Aug. 1984.; *Guardian*, 23 Nov. 1984. 'Science is not common-sense' was a comment made at a meeting of the British Association for the Advancement of Science.
14. 'Leukaemia Incidence in Somerset', Somerset Health Authority, County Hall, Taunton; Document S1889; see also *Ref.* 12.
15. Committee on Medical Aspects of Radiation in the Environment (COMARE), *Nature*, Vol. 329, pp. 499-505, 8 Oct. 1987; *Nature*, Vol. 342, p. 213, 16 Nov. 1989; see also WISE *News Communique*, No. 318, p. 7, 29 Sept. 1989.
16. *British Medical Journal*, Vol. 295, pp. 819-827, 3 Oct. 1987.
17. 'Britain links nuclear plants and leukaemia,' *Age*, 10 June 1988; see WISE *News Communique*, No. 295, p. 9, 1 July 1988.
18. Gardner M.J., *et al.*, 'Results of case control study of leukaemia and lymphoma among young people near Sellafield nuclear plant in West Cumbria', *British Medical Journal*, Vol. 300, pp. 423-9, 1990. See also 'Leukaemia sparks review of radiation limits', *New Scientist*, p. 4, 24 Feb. 1990.
19. Macht S. & Lawrence P., 'National Survey of Congenital Malformation Resulting from Exposure to Roentgen Radiation', *American Journal of Roentgenology*, Vol. 76, pp. 442-66, 1955.
20. McPhee, Letter, 'How radiation can affect sperm cells', *Age*, 5 Mar. 1990.
21. See *Ref.* 10.
22. 'Indecent Exposure', *Guardian*, 24 Nov. 1983.
23. *Chain Reaction*, No. 54, p. 12-13, Winter 1988, Friends of the Earth, Melbourne. Jean McSorley toured Australia for Greenpeace in May 1988. She has lived in Cumbria all her life. McSorley has been an active member of CORE — Cumbrians Opposed to a Radioactive Environment. She has taken part in studies and public inquiries on low-level ionising radiation and its health hazards. In 1990 Jean McSorley

came to Australia to work with Greenpeace on the uranium mining issue.

24. Shook L., 'The Hanford Report,' *Pan Environmental Journal*, Vol. 1, No. 1; 'How Dangerous is Hanford's Radiation? The Scientific Debate', p. 35-58, Pan Institute, 1720 West Fourth Ave., Spokane, USA, 1985. See also Wasserman H.& Solomon N., 'Nuclear Workers: Radiation on the Job', Ch 7, *Killing Our Own*, Delecorte Press, NY, 1982.
25. Ramey J.T., Commissioner, US Atomic Energy Commission in correspondence with Leo Goodman, Secretary, Atomic Energy Technical Committee, 19 September 1966.
26. US Congressional hearings on *Effect of Radiation on Human Health*, Hearings before the Subcommittee on Health and Environment of the Committee on Interstate and Foreign Commerce, House of Representatives, Vol. 1, p. 525. 1978, Government Printing Office, Washington DC.
27. See *Ref.* 24, p. 46.
28. Stewart A., *Transcript*, p. 6718.
29. Stewart A , *Transcript*, p. 6737.
30. Mancuso T.F., Stewart A.M, & Kneale G.K., 'Radiation Exposure of Hanford Workers Dying from Cancer and other Causes', *Health Physics*, Vol. 33, pp. 369-84, 1977.
31. Stewart A., *Transcript* p. 6728.

Chapter 11

1. Ford D., *Cult of the Atom — The Secret Papers of the Atomic Energy Commission*, p. 42, Simon & Schuster, NY, Revised Ed., 1984.
2. *Age*, 12 July 1986.
3. *The Accident at the Chernobyl Nuclear Power Plant and its Consequences: Information compiled for IAEA Experts' Meeting, 25-9 Aug. 1986*, Vienna I-III. Quoted by Haynes V. & Bojcun M., *The Chernobyl Disaster — The true story of a catastrophe — an unanswerable indictment of nuclear power*, pp. 1-18, Hogarth Press, London, 1988
4. Il'in L.A., 'The Chernobyl experience in the contemporary problems of radiation protection', *Proceedings of the Scientific Conference on the Medical Aspects of the Chernobyl Accident, Kiev, May, 1988*. Moscow Ministry of Public Health, 1988. Quoted by Medvedev, Z., *The Legacy of Chernobyl*, pp129-30, Basil Blackwell, Oxford, 1990.
5. Chaplain-Riou M., *Agence France Press*, 23 May 1989.
6. *Ecoropa Information Sheet* No. 15; quoted Quoted by Haynes, V. & Bojcun, M., *The Chernobyl Disaster*, p. 91.
7. World Information Service on Energy, *News Communique*, No. 330, p. 1, 6 April 1990.
8. *Ibid*, p. 2.
9. *Age*, 8 April 1990.
10. Gofman J., 'Assessing Chernobyl's Cancer Consequences: Application

of Four Laws of Radiation Carcinogenesis', 19th National Meeting, The American Chemical Society, 9 Sept. 1986. Available from Committee for Nuclear Responsibility, PO Box 11207, San Francisco, Ca 94101, USA. Gofman estimates 500,000 fatal cancers over the next 40 years in the Soviet Union and Western Europe. Another estimate for Western Europe is no measurable increase over the ordinary occurrence of cancer: 'Dose Distributions in Western Europe Following Chernobyl', by R.H. Clarke, UK National Radiological Protection Board in *Radiation and Health — The Biological Effects of Low-Level Exposure to Ionising Radiation*, in R. Russell Jones & R. Southwood (Eds), pp. 251-263, 1987; see also B.E. Lambert, 'The Effects of Chernobyl', Medical College, St Bartholomew's Hospital, London, who estimates 400-500 cancers in the UK. Monitoring of fallout at its peak was inadequate and so dose estimates will tend to be lower than those actually experienced. Also the official radiation bodies use the 1977 ICRP risk-estimates to assess the cancer fatalities and this is now acknowledged to be at least eight times too low.

11. World Information Service on Energy, *News Communique*, No. 290, April 1985. See also G. Luning, J. Sheer, M. Schmidt and H. Ziggel, 'Early infant mortality in West Germany before and after Chernobyl', *The Lancet*, pp. 1081-1083, 4 Nov. 1989. For a similar study in the United States see J.M. Gould and E.J. Sternglass, 'Low-level radiation and mortality', *Chemtech*, pp. 18-21, Jan. 1989.
12. *Guardian Weekly*, p. 2, 25 May 1986.
13. The European Commission (EC), the executive arm of the European Council of Ministers (ECM). recommends radioactivity levels in food, although legal authority still rests with member states. However the EC is claiming power under the Eurotom treaty to regulate food trade. The European Parliament (EP), which has only a consultative role, has debated the radioactive food issue. The majority of members, which includes Greens, want to set much lower levels. Blocking acceptance in the EP could mean a compromise or simply shelving the EC regulations. The EP will have more say after 1992. Information is available from Undine Von Blottnitz, Member European Parliament, Untergut, D-3131 Grabow, Federal Republic of Germany.
14. *Canberra Times*, 3 Jan. 1985, from Australian Archives cabinet papers; see also *Submission* to Senator Gareth Evans from the Sutherland Shire Residents Action Group on 'Review of the Atomic Energy Act with Reference to the Australian Atomic Energy Commission Research Establishment at Lucas Heights', 8 May 1985.
15. Moyal A., 'The Australian Atomic Energy Commission: A Case Study in Australian Science and Government', *Search*, Vol. 6, pp. 365-84, 1975.
16. See *Submission, Ref.* 14.
17. *Ibid*
18. Sutherland Shire Council in Committee, *Minutes*, 14 Feb. 1955.

19. Linacre E., 'Wind and Temperature in Sutherland Shire: A Preliminary Study Related to Air Pollution', Macquarie University School of Earth Sciences, Sydney.
20. Baston P., 'Notes on Strontium-90 in Milk Samples', 1908, see *Submission Ref.* 19.
21. See *Submission, Ref.* 14.
22. *Ibid.*
23. *Sydney Morning Herald*, 18 June 1986.
24. National Energy Research and Development Council, 1979; quoted in *Submission, Ref.* 19.

PART III

Chapter 12

1. Phillips J.L., Lecture given to students in Social Environmental Studies, Royal Melbourne Institute of Technology, June 1986. See also Phillips J.L., Winters W.D. & Rutledge L., '*In vitro* exposure to electromagnetic fields: changes in tumour cell properties', *International Journal Radiation Biology*, Vol. 49, pp. 463-9, 1986; 'Transferrin Binding to Two Human Colon Carcinoma Cell Lines: Characterisation', *Cancer Research*, Vol. 46, pp. 239-44, 1986.
2. New York State Powerline Project Scientific Advisory Panel, *Biological Effects of Powerline Fields, Final Report*, pp. 67-72, 1987. New York State Public Service Commission, July 1987.
3. Lyle D.B., Schechter P., Adey W.R. & Lundak R.L., *Bioelectromagnetics*, Vol. 4, pp. 281-92, 1983.
4. Dixey R. & Rein G., *Nature*, Vol. 296, pp. 253-6, 1982.
5. Adey W.R., 'Nerve Cell Membrane and Intracellular Communication in Brain Tissue', *Fourth Abbie Memorial Lecture*, University of Adelaide, 1985.
6. *Earthworm*, ABC, 6 May 1987.
7. Hamer J.R., 'Effects of low-level low-frequency electric fields on human reaction time,' *Communication Behavioural Biology*, Vol. 2, Part A, p. 217, 1968; Gavales-Medici R. & Day-Magdelino S.R. 'Extremely low-frequency weak electric fields affect schedule-controlled behaviour of monkeys', *Nature*, Vol. 261, p. 256, 1976.
8. Adey R., 'Tissue interactions with non-ionising electromagnetic fields', *Physiological Reviews*, Vol. 61, pp. 435-514, 1981.
9. See *Ref.* 6.
10. Nowell P.C. & Hungerford D.A. 'A minute chromosome in human chronic granulytic leukemia', *Science*, Vol. 132, p. 1497, 1960. Mitelman F., *et al.*, 'Non-random karyotypic evolution in chronic myeloid leukemia,' *International Journal Cancer*, Vol. 18, pp. 24-30, 1976; also Vol. 18, pp. 31-8, 1976.

11. Committee for the Compilation of Materials on Damage by the Atomic Bombings on Hiroshima and Nagasaki, *Hiroshima and Nagasaki — The Physical, Medical and Social Effects of the Atomic Bombings*, pp. 311-19, Hutchinson, London, 1979.
12. 'Protein Reveals Damage from Radiation', *New Scientist*, p. 41, 14 Jan. 1988.
13. Mueler H.J. 'Artificial transmission of the gene', *Science*, Vol. 66. p 84, 1927.
14. Wever R., 'ELF effects on human circadian rhythms', *ELF and VLF Electromagnetic Field Effects*, pp. 101-44; Plenum Press, New York, 1974. For a discussion of this work see Becker R.O. & Seldon G., *Body Electric*, pp. 248-9, William Morrow, New York, 1985.
15. See 'Biological Rhythms', *Ref.* 2, pp. 108-13.
16. Kavaliers M. & Ossenkopp K., 'Tolerance to morphine-induced analgesia in mice: Magnetic fields function as environmental specific cues and reduce tolerance development', *Life Science*, Vol. 37, pp. 1125-1135, 1985.
17. See *Ref.* 2, p. 111.

Chapter 13

1. Ball H., *Justice Downwind — America's Atomic Testing Program in the 1950s*', p. 113, Oxford University Press, NY, 1986.
2. McGinty L. & Atherly G., 'Acceptability versus Democracy', *New Scientist*, pp. 323-5, 1977 and Kletz T.A., 'What risks should we run?', *op.cit.* pp. 320-2. The former paper questions expert committees deciding an 'acceptable' risk while the latter paper discusses its application in the chemical industry.
3. Radford E.P., 'Recent evidence of radiation-induced cancer in Japanese atomic bomb survivors', *Radiation and Health — The biological effects of low-level exposure to ionizing radiation*, p. 92, Russell Jones R., & Southwood R.(Eds.), Wiley, Brisbane, 1987.
4. In answer to a question (No. 879, April 1988) in the House of Representatives on monitoring of imported food for radioactive contamination, the Minister for Science and Customs replied, 'Only a few items, such as hazelnuts and herbs have shown levels higher than 100 becquerels per kilogram but they have not posed a risk to health. They are often used in processed foods . . . and the concentration in the consumed food would then be very much less than 100 becquerels per kilogram.' Although a minor case of contamination, such dilution opens the way to disguising more serious contamination. The minister also indicated the Australian Government intended to adopt the proposed higher 'post Chernobyl' levels of radioactivity in food.
5. Modan B., *et al.*, 'Thyroid cancer following scalp irradiation', *Radiology*, Vol. 123, pp. 741-4, 1977.
6. See *Ref.* 3, p. 92.

7. Mancuso, T.F., Stewart, A.M, & Kneale, G.K., 'Radiation Exposure of Hanford Workers Dying from Cancer and other Causes', *Health Physics*, Vol. 33, pp. 369-84, 1977.
8. Stewart A. & Kneale G., 'Radiation Dose Effects in Relation to Obstetrics; X-ray and Child Cancer', *Lancet*, pp. 1185-7, 1970.
9. Gofman J.W., 'Protection by dose fractionation?', *Radiation and Human Health*, pp. 404-7. Sierra Club Books, San Francisco, 1981.
10. Knox E.G., Stewart A.M., Gilman E.A. & Kneale G.W., 'Background radiation and childhood cancer', *Journal of Radiology Protection*, Vol. 8, No1, pp. 9-18, 1988. See also 'Background radiation blamed for child cancers', *New Scientist*, 23 Oct. 1986.
11. Green P., 'Occupational Exposure Risks', *SCRAM Journal*, p. 14, Jan./Feb. 1987.
12. Committee on the Biological Effects of Ionising Radiation, *Health Effects of Exposure to Low Levels of Ionising Radiation: BEIR V*, National Academy Press, 1990.
13. ICRP 1990. Recommendations of the Commission: Draft Feb. 1990, ICRP/90/G-01 1990-02-09.
14. World Information Service on Energy, *News Communique* No. 329, p. 2, 9 Mar. 1990.
15. Green P., 'The response of the International Commission on Radiological Protection to calls for a reduction in the dose limits for radiation workers and members of the public', *International Journal of Radiation Biology*, Vol. 53, pp. 679-82, 1988.
16. 'Leukaemia sparks review of radiation limits', *New Scientist*, p. 4, 24 Feb. 1990.
17. Wagoner J. *et al*, 'Radiation as the cause of lung cancer among uranium miners', *New England Journal Medicine*, Vol. 273, pp. 181-8, 1965; see *Ref.* 11, 'Radon carcinogenesis: Lung cancer in the uranium miners', pp. 443-51.
18. Gofman J.W., *Radiation and Human Health*, pp. 790; Sierra Club Books, San Francisco, 1981.
19. Bertell R., 'Radiation and Heredity', *No Immediate Danger — Prognosis for a radioactive earth*, pp. 47, The Women's Press, London, 1985.

Index